Applied Mathematics

with Microsoft® Excel

Excel

Chester Piascik
Bryant College

BROOKS/COLE
™
THOMSON LEARNING

Australia • Canada • Mexico • Singapore • Spain • United Kingdom • United States

Assistant Editor: *Seema Atwal*
Marketing Manager: *Karin Sandberg*
Marketing Team: *Beth Kroenke*
Editorial Assistant: *Nathan Day*

Production Editor: *Rita Jaramillo*
Cover Design: *Vernon Boes*
Print Buyer: *Karen Hunt*
Printing and Binding: *Webcom*

For more information, contact
Wadsworth/Thomson Learning
10 Davis Drive
Belmont, CA 94002-3098
USA

For more information about our products, contact us: Thomson Learning Academic Resource Center
1-800-423-0563
http://www.wadsworth.com

International Headquarters
Thomson Learning
International Division
290 Harbor Drive, 2nd Floor
Stamford, CT 06902-7477
USA

UK/Europe/Middle East/South Africa
Thomson Learning
Berkshire House
168-173 High Holborn
London WC1V 7AA
United Kingdom

Asia
Thomson Learning
60 Albert Complex, #15-01
Singapore 189969

Canada
Nelson Thomson Learning
1120 Birchmount Road
Toronto, Ontario M1K 5G4
Canada

ISBN 0-534-37059-4

TABLE OF CONTENTS

PREFACE

This text uses Excel to serve the pedagogical needs of mathematics, not vice versa. It takes the viewpoint that the human mind is and must be superior to a machine, for it is the human mind with its will and intellect that results in the quest for good--that being the advancement of mankind. The mathematical development of the human mind is the paramount focus of this text. Accordingly, homework exercises are structured so that the student is encouraged to be an active participant in the learning process. This is accomplished by asking the student to verbalize the mathematical concepts behind spreadsheet results. Additionally, wherever feasible, the student is asked to verify spreadsheet results by applying underlying mathematical concepts. Finally, this text brings to the forefront the *thoughts between the lines* that professors hope their students have gained. This is accomplished by using the power of the spreadsheet to demonstrate useful relationships between quantities.

Audience
This text is written to serve the needs of college students enrolled in a college mathematics course for the managerial, life and social sciences. Moreover, the text envisions such students as needing additional assistance to gain a meaningful understanding of and appreciation for the usefulness of the subject matter. This text is written to provide that additional assistance through the use of Excel. The intention is to provide a text that teaches and uses Excel to clarify subject matter content and motivate students to learn such content by focusing on core concepts through a combination of lucid exposition and carefully crafted homework exercises that are both timely and relevant to today's business climate.

Why Excel?
The use of Excel to support the above-mentioned goals is justified because it is widely used in today's business marketplace. If not already doing so, today's students will be using Excel in other, nonmathematical courses. Furthermore, most professionals and students have access to Excel. For all of the above reasons, Excel is a natural tool to use in a mathematics course. The spreadsheets and illustrations in this text are mostly from Excel 97 and Windows 98. Although Excel instructions in this text are appropriate for Excel 97 and Windows 98, they are also appropriate for Excel 2000.

Features
All texts have an agenda. This one is no exception. In addition to the previously stated commentary, this text's agenda is defined by the following exposition of specific features.

Students need no prior experience with Excel. This text is designed so that professors do not have to use valuable class time to explain Excel instructions. Each section includes Excel instructions for relevant topics, and when needed, additional Excel instructions are provided within homework exercises. Furthermore, where appropriate, components of Excel formulas are explained so that the student understands the meaning of the formulas. In many cases, such explanations also reinforce underlying algebraic concepts. Moreover, Chapter 0, entitled Spreadsheet Basics, explains introductory spreadsheet topics and provides homework exercises containing discussions that, for example, explain relative versus absolute cell references along with Excel formulas and functions.

Lucid exposition complemented by motivating homework exercises. Each section begins with clear and efficient subject matter exposition followed by relevant spreadsheet instructions. This is complemented by homework exercises designed to illuminate the meaning behind important mathematical topics, reveal useful relationships, demonstrate the effects of change, develop intuition, and reinforce core concepts. A significant amount of learning takes place within the homework exercises. Where needed, some exercises include additional spreadsheet instructions. Exercises are designed to motivate students to study mathematics because it is meaningful to them and because its usefulness and relevance have been demonstrated through the use of Excel. In summary, mathematics and the understanding of mathematics, through the use of Excel, are the primary concerns of this text.

Integrates easily with other texts. Textbook contents are keyed to tables of contents of leading texts on college mathematics for the management, life, and social sciences.

Emphasis on use, meaning, and intuitive understanding of mathematical concepts. The interpretation of slope is enhanced by spreadsheet exercises involving cost functions, simple interest (linear growth), inventory functions (linear decay), and depreciation (linear decay). Spreadsheet exercises illuminate the meaning of present value of an annuity by creating a table illustrating periodic withdrawals from an account where the present value of the annuity is the initial balance. The **Linear Programming** chapter contains exercises where students are required to substitute the solution into constraint equations in order to verify their feasibility and slack or surplus values. Also, students are required to substitute solution values into objective functions to verify spreadsheet results.

Pencil and paper homework exercises ask students to verbalize mathematical concepts. These are included to encourage students to interact with and actively engage the subject matter. Interjected at appropriate points in homework exercise sets, such exercises ask students to summarize recently learned concepts, verify spreadsheet results, verbalize recently demonstrated relationships, or comment on the results of previous spreadsheet exercises.

Easy to use by both professor and student. In addition to designing the text so that professors do not have to use valuable class time to explain Excel instructions, we have used headings to reveal the themes behind certain homework exercises. Such headings allow both students and professors to identify quickly the issues involved in such exercises.

Includes timely and relevant topics not found in other texts. Notable examples of this feature include Chapter 0, **Spreadsheet Basics**, where the discussion centers on currency exchange rates, and spreadsheet fundamentals are simultaneously presented. Consequently, these currency exchange concepts are used in the section on mathematics of finance entitled, **International Investing**, where the effect of currency appreciation or depreciation on investment returns is demonstrated. Today's global economy demands such knowledge for intelligent decision making. Another example of this feature occurs in a section on **Amortization** that provides spreadsheet exercises on *mortgage reduction* where, for example, the effect of making an extra $1000 payment is easily determined. A section on **Automobile Leases** includes spreadsheet exercises that illustrate how the capitalized cost affects the monthly lease payment, how the residual value affects the monthly lease payment, and other topics.

Enriched spreadsheet exercises illuminate dynamics of change. Specifically, the normal distribution section contains exercises on generating a bimodal distribution, on simulating the mixing of output from two production processes, on overcontrol of a production process, and other topics. Another section provides spreadsheet exercises where profit functions illustrate how a $1 decrease in the fixed cost increases profit by $1. This concept is applicable to the downsizing of companies that took place during the early nineties to propel the stock market to new highs. A separate chapter prepares students for a more meaningful understanding of definite integrals by beginning with rate-of-change step functions, such as velocity functions, and demonstrating intuitively that the area under the graph of a velocity function gives the total distance traveled during the time interval. This idea is generalized to develop the concept that the *area under the graph of y' gives the total change in y over the interval*. The following section builds on this presentation to develop the concept of Riemann sums in a more meaningful manner.

Enhanced pedagogy not found in other texts. A separate chapter, Chapter 2, is devoted to graphs of functions that are to appear in the calculus chapters. This chapter provides detailed discussions and spreadsheet exercises on power functions, along with the concepts of vertical and horizontal shifts and reflections in the x-axis. A separate section compares graphs of quadratic functions of the form $y = ax^2 + bx$ and of the form

$y = ax^2 + bx + c$. The intent is to develop the power to graph such functions before the encounter with calculus so that, during the calculus chapters, *the student can focus on the calculus concepts* instead of wondering how a particular graph was obtained from a certain equation. Concepts of rate of change and acceleration are made more concrete through the use of discrete data and spreadsheets by computing y'-values as the difference between successive y-values and y''-values as the difference between successive y'-values. The concept that acceleration is the rate of change of a rate of change becomes more meaningful, and the stage is set for greater intuitive understanding of the second derivative. The section on exponential growth uses spreadsheets to demonstrate that for larger x-values, exponential growth outpaces polynomial growth. Analogously, a section on logarithmic functions uses spreadsheets to demonstrate that for larger x-values, polynomial growth outpaces logarithmic growth. The power of the spreadsheet is used to demonstrate that a graph of ln y versus x results in a straight line if y versus x exhibits exponential growth or decay.

ACKNOWLEDGMENTS

First I thank my colleagues at Bryant College for honest discussions and comments freely given during our department meetings and one-on-one conversations. Along with my classroom experience, these exchanges have helped shape some of the ideas behind the development of this text. I consider myself fortunate to have this group of people for colleagues. My colleagues are Nancy Beausoleil, James Bishop, Marcia Gee, Louise Hasenfus, Kristin Kennedy, Kunio Mitsuma, Robert Muksian, Patricia Odell, Alan Olinsky, John Quinn, Martin Rosenzweig, Phyllis Schumacher, and Richard Smith.

I thank my editor, Stephanie Schmidt, at Brooks/Cole Publishing for her dedicated efforts in the production of this text. I thank my production coordinator, Stephanie Andersen, for guiding this text through the production process. Also, I thank Seema Atwal for her efforts during the early phase of this project.

I thank the reviewers for helpful comments and perceptive suggestions, many of which have been implemented in this text. They are Cami Bates at the University of Denver, John Nelson at Lane Community College, Tom Obremski at the University of Denver, and Stefan Waner at Hofstra University.

CHAPTER ZERO

Spreadsheet Basics

Spreadsheet 0-1 gives *currency exchange rates* between the U.S. dollar and currencies listed in Column A. Notice that spreadsheet cells are identified by a column letter and row number.   For example, the upper left cell is identified as cell A1; the cell containing the word "yen" is cell A3; the cell containing the number 130 is cell C3.

SPREADSHEET 0-1

	A	B	C	D	E	F	G	H
1		Currency per U S $				U S $ equivalent		
2		E	L	%Ch		1/E	1/L	%Ch
3	yen	117	130					
4	lira	1890	1650					
5	peso	9.4	9.2					
6	rand	5.5	6.2					

We take this opportunity to explain useful concepts about currency exchange rates. The left portion of Spreadsheet 0-1 labeled "Currency per U S $" gives the value of the U S dollar in terms of the indicated foreign currency both at early (*E*) and late (*L*) points in time. For example, at some point in time designated as early (*E*), the exchange rate is given as 117. This means that at this *early point* in time,

$$\$1 \text{ buys } 117 \text{ yen} \text{ or, equivalently, } \$1 = 117 \text{ yen}.$$

The exchange rate 130 means that at the *later point* in time,

$$\$1 \text{ buys } 130 \text{ yen} \text{ or, equivalently, } \$1 = 130 \text{ yen}.$$

Now, you should create Spreadsheet 0-1 on your computer using the following instructions.

INSTRUCTIONS

Use the following instructions to create tables similar to those in Spreadsheet 0-1.

1. Open Excel to get a blank worksheet. If a blank worksheet does not appear, create a new worksheet by selecting **File** from the menu bar and then selecting **New**.

2. Use the mouse to move your pointer to cell B1 and click the left mouse button to make that cell the active cell. A rectangle with a dark border should appear around cell B1.

Type the label **Currency per U.S. $**. Note that the label continues to adjacent cells until you stop typing.

3. Continue typing labels and numbers until you have created Spreadsheet 0-1 on your spreadsheet.

INSTRUCTIONS

Use the following instructions to **ENTER** and **COPY** a **FORMULA.**

Next, we use the spreadsheet to determine the *rate of appreciation or depreciation* of the U. S. dollar against the indicated currencies by using the formula for percent change,

$$\frac{LateValue - EarlyValue}{EarlyValue} .$$

1. We begin by showing how to *enter a formula* with Excel. Use your mouse to move the dark-bordered rectangle to cell D3, type the formula **=(C3-B3)/B3**, and press **Enter**. This enters the formula for percent change given above. Note that with Excel, *formulas are always preceded by an equals (=) sign.* Also note that the entered formula, **=(C3-B3)/B3**, computes the percent change (130 -117)/117, which equals 0.11111, or 11.1%, of the dollar against the yen. Observe that 0.111111 appears in cell D3, which means that the dollar has appreciated 11.11% against the yen. In other words, the dollar buys 11.11% more yen now than at the earlier point in time. Later, we'll show how to change decimal results (0.111111) to percents on the spreadsheet.

2. Now, we must *copy the formula* down through cell D6 to get the percent changes in the value of the dollar against the remaining listed currencies. Make certain that the dark-bordered rectangle is at cell D3. Use the mouse to move the pointer to the small black box (called a handle) at the lower right-hand corner of cell D3. The mouse pointer becomes a thick black plus. Click the mouse button without releasing it and drag the mouse pointer down to cell D6. Release the mouse button at cell D6, and cells D3 through D6 will contain the formula values (i.e., the percent changes) for the remaining currencies. Your spreadsheet should resemble Spreadsheet 0-2.

SPREADSHEET 0-2

	A	B	C	D	E	F	G	H
1		Currency per U S $				U S $ equivalent		
2		E	L	%Ch		1/E	1/L	%Ch
3	yen	117	130	0.111111				
4	lira	1890	1650	-0.126984				
5	peso	9.4	9.2	-0.021277				
6	rand	5.5	6.2	0.127273				

The percent change, -0.126984, in cell D4 means that during the period between the early and late points in time, the dollar has depreciated by approximately 12.7% against the lira. In other words, the dollar now buys 12.7% less lira now than at the earlier point in time.

INSTRUCTIONS

Use the following instructions to **change cell entries to percents**.

Specifically, we show how to change the contents of cells D3 through D6 to percents.
1.   Move the mouse pointer to the middle of cell D3 and click until a thick white cross appears in the middle of cell D3. Hold and drag the mouse pointer down through cell D6. Cell D3 will remain unhighlighted while cells D4 through D6 are highlighted in black.

2.   Select **Format** from the menu bar and click on **Cells**. Click on **Percentage** in the Category text box. The white box at the right should show **two** decimal places. If needed, click on the up arrow or down arrow to select the required number of decimal places. Click **OK** and your spreadsheet should resemble that of Spreadsheet 0-3.

SPREADSHEET 0-3

	A	B	C	D	E	F	G	H
1		Currency per U S $				U S $ equivalent		
2		E	L	%Ch		1/E	1/L	%Ch
3	yen	117	130	11.11%				
4	lira	1890	1650	-12.70%				
5	peso	9.4	9.2	-2.13%				
6	rand	5.5	6.2	12.73%				

INSTRUCTIONS

Use the following instructions to **compute reciprocals**.

Specifically, we fill in the right portion of the spreadsheet, labeled **U S $ equivalent**, to get the currency exchange rates per unit of the foreign currency. Here, each cell entry gives the value of one unit of the foreign currency in terms of U. S. dollars and is therefore the *reciprocal* of the corresponding cell entry from the left portion of the spreadsheet. This is why the columns are labeled **1/E** and **1/L**, respectively.

1.   Move to cell F3 and type the formula **=1/B3** and press **Enter** to get the reciprocal of cell B3 contents. *Copy* this formula down through cell F6 by using the mouse to move the pointer to the small black box (called a handle) at the lower right-hand corner of cell F3. The mouse pointer becomes a thick black plus. Click the mouse button without releasing it and drag the mouse pointer down to cell F6. Release the mouse button at cell F6, and cells F3 through F6 will contain the reciprocals of cells B3 through B6.

2.   Repeat Step 1 for the **1/L** entries of cells G3 through G6.

3.   Compute the *percent changes* in cells H3 through H6 in the same manner that we computed those of cells C3 through C6 and change the cell entries to percents. Your spreadsheet should resemble Spreadsheet 0-4.

	A	B	C	D	E	F	G	H
1		Currency per U S $				U S $ equivalent		
2		E	L	%Ch		1/E	1/L	%Ch
3	yen	117	130	11.11%		0.008547	0.007692	-10.00%
4	lira	1890	1650	-12.70%		0.000529	0.000606	14.55%
5	peso	9.4	9.2	-2.13%		0.106383	0.108696	2.17%
6	rand	5.5	6.2	12.73%		0.181818	0.16129	-11.29%

The exchange rate, 0.008547, of cell F3 means that at the **early point** in time,

1 yen buys $0.008547 or, equivalently, *1 yen = $0.008547*.

The exchange rate, 0.007692, of cell G3 means that at the **later point** in time,

1 yen buys $0.007692 or, equivalently, *1 yen = $0.007692*.

The percent change, -10.00%, of cell H3 means that during the period between the early and later points in time, the yen has depreciated by 10% against the U. S. dollar. In other words, the yen now buys 10% less dollars now than at the earlier point in time.

EXERCISES

1. *Currency exchange rates.* Spreadsheet 0-5 gives currency exchange rates between the U.S. dollar and currencies listed in Column A.

SPREADSHEET 0-5

	A	B	C	D	E	F	G	H
1		Currency per U S $				U S $ equivalent		
2		E	L	%Ch		1/E	1/L	%Ch
3	peso	0.99	0.98					
4	real	1.7695	1.752					
5	krone	7.198	7.17					
6	guilder	2.1333	2.1587					

(a) Complete Spreadsheet 0-5 so that it reveals information comparable to that of Spreadsheet 0-4. Save your results by clicking **File** on the menu bar and then selecting **Save As**.

(b) Pencil and Paper Exercise. Interpret the exchange rates 1.7695 and 1.752 of cells B4 and C4.

(c) Pencil and Paper Exercise. Interpret the exchange rates 7.198 and 7.17 of cells B5 and C5.

(d) Pencil and Paper Exercise. State and interpret the percent change of cell D3.

(e) Pencil and Paper Exercise. State and interpret the percent change of cell D6.

(f) Pencil and Paper Exercise. Interpret the exchange rates of cells F4 and G4.

(g) Pencil and Paper Exercise. Interpret the exchange rates of cells F6 and G6.

(h) <u>Pencil and Paper Exercise</u>. State and interpret the percent change of cell H5.

(i) <u>Pencil and Paper Exercise</u>. State and interpret the percent change of cell H6.

(j) Move to cell B3 and change the 0.99 to 0.93 and note how resulting spreadsheet entries are updated.

(k) Move to cell B5 and change the 7.198 to 7.10 and note how resulting spreadsheet entries are updated.

2. *Clearing a spreadsheet.* Assume you have saved Spreadsheet 0-5 and wish to clear it prior to proceeding to Exercise 3. Move the mouse pointer to the middle of cell A1 and click until a thick white cross appears in the middle of cell A1. Hold and drag the mouse pointer across and down until the highlighted region includes all that you want to clear. Release the mouse, and a rectangular region containing the cells that you designated to clear will be highlighted in black. The original cell, in this case A1, will remain unhighlighted.

3. *Relative versus absolute cell references.* Spreadsheet 0-6 illustrates the difference between *relative* and *absolute* cell references.

SPREADSHEET 0-6

	A	B	C	D	E	F
1	Relative Cell Reference			Absolute Cell Reference		
2		"=5*A4"		"=5*A$4"		
3	x	y		y		
4	0	0		0		
5	1	5		0		
6	2	10		0		
7	3	15		0		
8						
9	Relative Cell Reference					
10		x	0	1	2	3
11	"=5*C10"	y	0	5	10	15
12						
13						
14	Absolute Cell Reference					
15		x	0	1	2	3
16	=5*$C10"	y	0	0	0	0

The upper portion of Spreadsheet 0-6 illustrates the creation of a table of *x*- and *y*-values for the formula $y = 5x$. We will use this formula to demonstrate the difference between relative and absolute cell references.

(a) Create the upper left-hand corner of Spreadsheet 0-6, labeled **Relative Cell Reference**, on your computer as follows. After entering the labels and *x*-values 0, 1, 2, 3, move to cell B4 and *enter the formula* =5*A4 and *copy the formula* down through cell B7. The * denotes multiplication. Your result should contain the *y*-values, 0, 5, 10, and 15, shown in the spreadsheet. Each *y*-value was obtained by multiplying the indicated *x*-value by 5. As the formula =5*A4 was copied down through cell B7 (in other words, as 5 was multiplied by the respective *x*-values in column A), it was applied to cells A5

5

through A7, which means that the cell reference to A4 kept changing. This is what is meant by a *relative* cell reference.

(b) Now, we demonstrate an *absolute* cell reference. Move to cell D4 and enter the formula =5*A$4 and copy the formula down through cell D7. The dollar sign $ in front of the row reference, **4**, means that the reference to Row 4 is to remain *fixed* or *constant* and does not change as the formula is copied down through cell D7. The result is that the *y*-values do not change but remain at the first *y*-value of 0.

(c) The bottom portion of Spreadsheet 0-6 demonstrates an *absolute* cell reference where the dollar sign $ appears in front of the column reference. First, note the bottom section below the label **Relative Cell Reference** beginning at cell B10, where the label *x* appears followed by the *x*-values 0, 1, 2, and 3 listed horizontally through cell F10. Just below, in cell C11, we enter the formula =5*C10 and copy it horizontally through cell F11 to get the *y*-values 0, 5, 10, and 15. Of course, this is an example of a *relative cell reference*. Create this on your spreadsheet now.

Next, note the section below labeled **Absolute Cell Reference** where we see the same *x*-values along with *y*-values generated by the formula =5*$C10. Here, the dollar sign $ in front of the column reference, **C**, means that the reference to Column C is to remain *fixed* or *constant* and does not change as the formula is copied down through cell D7. The result is that the *y*-values do not change but remain at the first *y*-value of 0. This is another example of an *absolute cell reference.* Create this on your spreadsheet now.

4. *Getting acquainted: Multiplication takes precedence over addition and subtraction.*
(a) Move to an empty cell and type the formula =5+7*2. Press **Enter**, and the result, 19, should appear in the cell. Explain how multiplication takes precedence over addition and subtraction.
(b) Move to an empty cell and type the formula =8 -2*3. Press **Enter**, and the result, 2, should appear in the cell. Explain how multiplication takes precedence over addition and subtraction.

5. *Getting acquainted: Left-to-right evaluation.*
(a) Move to an empty cell and type the formula =15/5*2. Press **Enter**, and the result, 6, should appear in the cell. Because multiplication and division have the same priority level, Excel evaluates the formula from left to right. Explain how Excel determined the result, 6.
(b) Move to an empty cell and type the formula =10*3/2. Press **Enter**, and the result, 15, should appear in the cell. Because multiplication and division have the same priority level, Excel evaluates the formula from left to right. Explain how Excel determined the result, 15.

6. *Getting acquainted: Exponentiation, ^, takes precedence over multiplication and division.*
(a) Move to an empty cell and type the formula =5+7*2^3. Press **Enter**, and the result, 61, should appear in the cell. Note that the symbol ^ denotes exponentiation which takes precedence over multiplication and division. Explain how Excel determined the result, 61.
(b) Move to an empty cell and type the formula =8+5*3^2. Press **Enter**, and the result, 53, should appear in the cell. Explain how Excel determined the result, 53.

7. *Getting acquainted: Implied multiplier of -1.*

(a) Use Excel to evaluate -4^2 by moving to an empty cell and typing the formula **=-4^2**. Press **Enter**, and the erroneous result, 16, appears in the cell. The result, 16, is wrong because the expression -4^2 implies that only the 4 should be squared, not the negative sign. Thus, the correct answer is -16. To avoid this error, use the implied multiplier **-1** by entering the formula **=-1*4^2**. The result is -16.

(b) Use Excel to evaluate -5^2.

(c) Generate a table of x- and y-values for the equation $y = -x^2$ as follows. Clear the spreadsheet, move to cell A1 and enter x, move to cell B2 and enter y. Enter 1, 2, 3, 4 in cells A2 through A5, respectively. Move to cell B2 and enter the formula **=-1*A2^2** and *copy* the formula down through cell B5. Explain why we used the multiplier, *-1*, in the formula.

8. *Getting acquainted: SUM function.*

(a) Move to cell A3 and type the formula **=SUM(A1:A2)** and press **Enter**. This command (also called the SUM function) gives, in cell A3, the sum of the numbers in cells A1 through A2. Here, the number 80 (i.e., $19 + 61$) should appear in cell A3.

(b) Another way of finding a sum is to use the **AutoSum tool**. Move to cell C5 and type the numbers 5, 6, 7, 8 in cells C5 through C8, respectively. Move the dark-bordered rectangle to cell C9 and click $\sum$ on the toolbar above your spreadsheet. Observe that Excel guesses the cells, C5 through C8, that you wish to sum by including them in the SUM formula. Press **Enter** or click $\sum$ and the sum, 26, appears in cell C9.

9. *Getting acquainted: AVERAGE, SQRT, EXP, LN functions.*

(a) Type the numbers 4, 5, 6, 7, 8 in cells D3 through D7, respectively. Move to cell D8 and type the formula **=AVERAGE(D3:D7)**. Press **Enter**, and the average of the numbers in cells D3 through D7 appears in cell D8.

(b) Move to any empty cell, type the formula **=SQRT(49)**, and press **Enter**, and the square root, 7, appears in the cell. Use this function to find $\sqrt{36}$.

(c) Move to any empty cell, type the formula **=EXP(3)**, and press **Enter**, and the value of e^3, 20.08554, appears in the cell. Use this function to find e^5.

(d) Move to any empty cell, type the formula **=LN(8)**, and press **Enter**, and the value of $\ln 8$, 2.079442, appears in the cell. Use this function to find $\ln 30$.

Note: The above exercises have demonstrated only a few of the many Excel functions available. Although we have used these functions by typing their formulas manually, they can also be used by clicking f_x on the toolbar above the spreadsheet and selecting them from the list of the many functions available with Excel. These include AVERAGE, SUM, SQRT, EXP, and LN that we have entered manually.

10. *Getting acquainted: Fractional exponents.*

(a) Although square roots can be determined by using the SQRT function, they can also be determined by using *fractional exponents*. For example, we use Excel to determine $\sqrt{49}$ by moving to any empty cell and entering the formula **=49^(1/2)**. Demonstrate that the parentheses are needed around the fractional exponent by moving to an empty cell and entering the formula =49^1/2. The erroneous result, 24.5, appears in the cell. Explain how Excel obtained this result and how including the parentheses corrects the mistake.

(b) Use fractional exponents with Excel to determine the values of $\sqrt{64}$, $\sqrt{81}$, and $\sqrt{16}$.

11. ***Another formula for computing percent change.*** In this chapter, we have used the

formula $\dfrac{LateValue - EarlyValue}{EarlyValue}$ to compute percent change.

(a) <u>Paper and Pencil Exercise</u>. Use algebra to show that this formula is equivalent to the

formula $\dfrac{LateValue}{EarlyValue} - 1$.

(b) <u>Paper and Pencil Exercise</u>. A stock's price goes from $24 to $30 during some time period. Use both formulas to compute the percent change in the stock's price. Remember to move the decimal point 2 places to the right to change the answer to a percent.
(c) <u>Paper and Pencil Exercise</u>. A stock's price goes from $50 to $20 during some time period. Use both formulas to compute the percent change in the stock's price. Remember to move the decimal point 2 places to the right to change the answer to a percent.

12. ***Error Messages.*** The following is a partial list of Excel error messages and what they mean:

#DIV/0!	Attempted division by 0
#NUM!	There is a problem with a number
#REF!	There is an invalid cell reference

Note: If ######## appears in a cell, this means that the number, as formatted, does not fit in the cell. One can ***widen a cell*** by moving the mouse pointer to the column heading at the top of the spreadsheet so that it points to the *right-hand border* of the *column*. The mouse pointer becomes a two-way arrow. *Double-click the right-hand border of the column*, and the number should appear in the cell.

CHAPTER ONE

Linear Functions

1-1 Slope–Intercept Form

The *slope–intercept* form of a linear equation is given by

$$y = mx + b$$

where m is the *slope* and b is the *y-intercept* of the corresponding straight line. This section is designed to enhance your understanding of the slope–intercept form. As an example, we consider the linear equation

$$y = 5x + 10.$$

Spreadsheet 1-1 gives a table of x- and y-values of this linear equation, along with its corresponding graph.

SPREADSHEET 1-1

x	y=5x+10
0	10
1	15
2	20
3	25
4	30
5	35
6	40

Studying the table and graph of Spreadsheet 1-1, note that the y-intercept, $y = 10$, corresponds to $x = 0$. To understand the concept of *slope*, we place a pointer or pencil at the y-intercept on the graph and move upward along the straight line towards the point $(6, 40)$. Note that as we move up the line, both x- and y-values are changing simultaneously such that *for every unit increase in x, the y-value increases by 5 units.* We confirm this by observing the *y*-values in either the table or the graph and noting that *successive y-values increase by 5 as the x-values increase by 1.* This is what we mean when we say that

1. *Slope* is the rate of change of *y* with respect to *x*.
2. *Slope* is the ratio of the change in *y* to the change in *x*.
3. *Slope* gives the effect upon *y* of a one-unit change in *x*.
4. *Slope* indicates the steepness or pitch of a straight line.

INSTRUCTIONS

Use the following instructions to create tables and graphs similar to those in Spreadsheet 1-1.

1. Type Labels
1.1 As an example, we will create a table of *x*- and *y*-values for the function featured in Spreadsheet 1-1. Thus, we will use Column A for the *x*-values and Column B for the *y*-values of the equation $y = 5x + 10$.

1.2 Once you have a blank worksheet, use the mouse to move your pointer to cell A1 and click the left mouse button to make that cell the active cell. A rectangle with a dark border should appear around cell A1. Use the spacebar to move to the middle of the cell and type **x**. This labels Column A. Now, we will label Column B with its equation. Use either the mouse or the arrows on the keyboard to move the dark-bordered rectangle around cell B1. Type **y=5x+10** in cell B1.

2. Create a Table of *x*- and *y*-values
2.1 Beginning with cell A2, type the *x*-values 0 through 6 into Column A as illustrated in Spreadsheet 1-1.

2.2 Now, we show how to *enter a formula* to compute the corresponding y-values for the equation $y = 5x + 10$ in Column B. Move the dark-bordered rectangle to cell B2 and type the formula **=5*A2+10** and press **Enter**. Note that the symbol ***** means multiplication. The formula **=5*A2+10** entered in cell B2 returns to cell B2 the *y*-value (*y* = 10) corresponding to the *x*-value (*x* = 0) of cell A2.

2.3 Now, we *copy the formula* down through cell B7. Move the dark-bordered rectangle to cell B2. Use the mouse to move the pointer to the small black box (called a handle) at the lower right corner of cell B2. The mouse pointer becomes a thick black plus. Click the mouse button without releasing it and drag the mouse pointer down to cell B7. Release the mouse button at cell B7, and cells B3 to B7 will contain the formula values—in other words, the *y*-values corresponding to the *x*-values of Column A.

10

3. Create a Graph of the Data

3.1 Move the mouse pointer to the middle of cell A1, click, hold, and drag the mouse pointer until it highlights the cells containing the data and labels. Cell A1 will remain unhighlighted.

3.2 Select **Chart Wizard** from the toolbar, and a dialog box appears.

Step 1: Select **XY(Scatter)** in the Chart type section. Move to the Chart sub-type section and click on the second graph in the first column. Click **Next**.

Step 2: Click **Next**.

Step 3: Enter **x** in the Value(X) axis section and **y** in the Value(Y) axis section to label the x- and y-axes. Click **Next**.

Step 4: Click **Finish** and the graph should appear in your spreadsheet.

4. Change the Size of the Graph

Move the mouse pointer to the middle handle at the bottom of the chart until the pointer becomes a vertical double-sided arrow. Click, hold, and drag the line downward to enlarge the chart vertically. To enlarge the chart horizontally, move the mouse pointer to the middle handle at the side of the chart until the pointer becomes a horizontal double-sided arrow. Click, hold, and drag the line horizontally to enlarge the chart. Dragging the line horizontally in the reverse direction will decrease the size of the chart.

5. A Faster Way of Creating a Table of x- and y-values

5.1 After typing **x** and **y=5x+10** in cells A1 and B1, respectively, type **0** in cell A2 and **1** in cell A3. Move to cell B2 and enter the formula **=5*A2+10**. Copy the formula down to cell B3 by moving the dark-bordered rectangle to cell B2. Then use the mouse to move the pointer to the small black box (called a handle) at the lower right corner of cell B2. The mouse pointer becomes a thick black plus. Click the mouse button without releasing it and drag the mouse pointer down to cell B3. Release the mouse button at cell B3, and cell B3 will contain the formula values.

5.2 To create the rest of the table of x- and y-values, move the mouse pointer to the middle of cell A2, click, hold, and drag the mouse pointer to the right so that it highlights cell B2 and then down to cell B7. Cells A3 through B8 should be highlighted with cell A1 remaining unhighlighted.

5.3 Then, from the Menu bar, choose **Edit, Fill, Series** commands. In the dialog box that appears, make certain that **Series in Columns** is selected in the **Type** box. Finally, select the **AutoFill** option and click **OK**. The column of x- and y-values should be complete.

5.4 Now, continue with the procedure used to Create a Graph of the Data.

EXERCISES

1. Create a table of x- and y-values for the equation $y = 4x + 7$.

   (a) Use the x-values 0, 1, 2, 3, 4, and 5.
   (b) Use Chart Wizard to create the corresponding graph.
   (c) Observing the result of part (b), state the value of the y-intercept.
   (d) *Slope interpretation.* Studying the table and graph for parts (a) and (b), state what's happening to the y-values as the x-values increase by 1.
   (e) *Slope interpretation.* Studying the table and graph for parts (a) and (b), state what's happening to the y-values as the x-values decrease by 1.

2. Create a table of x- and y-values for the equation $y = -2x + 3$.

   (a) Use the x-values 0, 1, 2, 3, 4, and 5.
   (b) Use Chart Wizard to create the corresponding graph.
   (c) Observing the result of part (b), state the value of the y-intercept.
   (d) *Slope interpretation.* Studying the table and graph for parts (a) and (b), state what's happening to the y-values as the x-values increase by 1.
   (e) *Slope interpretation.* Studying the table and graph for parts (a) and (b), state what's happening to the y-values as the x-values decrease by 1.

3. Pencil and Paper Exercise. *(Significance of the sign of the slope)*
   Draw a straight line that has: (a) Positive slope   (b) Negative slope

   Summary. Straight lines with positive slopes slant *upward to the right*, whereas straight lines with negative slopes slant _____*to the right*.

4. Create a table of x- and y-values for the equations $y = 4x + 3$, $y = 4x$, and $y = 4x - 3$.

   (a) Use the x-values 0, 1, 2, 3, 4, and 5. Also, use Column A for the x-values, Column B for the y-values of $y = 4x + 3$, Column C for the y-values of $y = 4x$, and Column D for the y-values of $y = 4x - 3$. Remember to highlight all four columns when creating the graphs.
   (b) Use Chart Wizard to create the corresponding graph.
   (c) Straight lines having the same slope are said to be *parallel*. Observing the result of part (b), note that the three lines are parallel. State the value of the y-intercept for each straight line.
   (d) Still observing the result of part (b), write the equation of the straight line that passes through the origin, (0, 0).

5. Pencil and Paper Exercise. Sketch the graphs of $y = -2x + 5$, $y = -2x$, and $y = -2x - 5$ on the same set of axes. Label the y-intercept of each line with its y-coordinate.

6. Pencil and Paper Exercise. Sketch the graphs of $y = 2x$ and $y = 5x$ on the same set of axes. State which line is steeper and explain why.

7. <u>Pencil and Paper Exercise</u>. Sketch the graphs of $y = 3x + 1$ and $y = 7x + 1$ on the same set of axes. State which line is steeper and explain why.

8. *Cost function.* The total cost in dollars, y, of producing x units of some product is given by

$$y = 10x + 30.$$

(a) Use the x-values 0, 1, 2, 3, 4, and 5 to create a table of x- and y-values.
(b) Use Chart Wizard to create the corresponding graph.
(c) Observing the result of part (b), state the value of the y-intercept.
(d) For a cost function, the y-intercept is also called the _____.
(e) Explain the meaning of the answer to part (d) in terms of *cost*.
(f) *Slope interpretation.* Studying the table and graph for parts (a) and (b), state what's happening to the y-values as the x-values increase by 1.
(g) Explain the meaning of the answer to part (f) in terms of *cost*.

9. *Simple interest: Linear growth.* If $2000 earns simple interest at an annual rate of 4%, then the interest for one year is determined by multiplying $2000 by 4% or, equivalently,

$$(2000)(0.04) = \$80.$$

This means that the $2000 investment increases in value by $80 per year. If y denotes this investment's value after x years, then the equation that relates y and x is
$y = 2000 + 80x$.

(a) Use the x-values 0, 1, 2, 3, 4, and 5 to create a table of x- and y-values.
(b) Use Chart Wizard to create the corresponding graph.
(c) Observing the result of part (b), state the value of the y-intercept. _____ In this example, the y-intercept is called the *initial investment.*
(d) *Slope interpretation.* Studying the table and graph for parts (a) and (b), state what's happening to the y-values as the x-values increase by 1.
(e) Explain the meaning of the answer to part (d) in terms of *investment value.*

<u>Summary</u>. If a linear equation with *positive slope* relates y with x, where x *denotes time*, then the y-values *increase* by the amount of the slope per unit of time and are said to exhibit *linear growth*. Thus, in this example, the investment's value is *growing* with the progression of time.

10. *Simple interest: Linear growth.* An investment of $8000 earns simple interest at an annual rate of 10%.
(a) Compute the interest earned in 1 year.
(b) The $8000 investment increases in value by $_____ per year.
(c) If y denotes this investment's value after x years, write the equation that relates y and x.
(d) Use the x-values 0, 1, 2, 3, 4, and 5 to create a table of x- and y-values.
(e) Use Chart Wizard to create the corresponding graph.
(f) Observing the result of part (e), state the value of the y-intercept. _____ In this example, the y-intercept is called the _____.

13

(g) *Slope interpretation*. Studying the table and graph for parts (d) and (e), state what's happening to the y-values as the x-values increase by 1.
(h) Explain the meaning of the answer to part (g) in terms of *investment value*.

11. ***Inventory function: Linear decay***. A product has an initial inventory level of 100 units, which decreases at the rate of 20 units per day. If y denotes the inventory level of this product after x days, the linear equation that relates y and x is $y = 100 - 20x$.
   (a) Use the x-values 0, 1, 2, 3, 4, and 5 to create a table of x- and y-values.
   (b) Use Chart Wizard to create the corresponding graph.
   (c) Observing the result of part (b), state the value of the y-intercept. _____ For an inventory function, the y-intercept is also called the *initial inventory*.
   (d) *Slope interpretation*. Studying the table and graph for parts (a) and (b), state what's happening to the y-values as the x-values increase by 1.
   (e) Explain the meaning of the answer to part (d) in terms of *inventory level*.

   Summary. If a linear equation with *negative slope* relates y with x, where x *denotes time*, then the y-values *decrease* by the amount of the slope per unit of time and are said to exhibit *linear decay*. Thus, in this example, inventory level is *decaying* with the progression of time.

12. ***Simple depreciation: Linear decay***. A grocery chain buys a new freezer for $30,000. If the freezer depreciates by $5000 per year, the equation that gives its value, y (in $thousands), after x years is $y = 30 - 5x$.
   (a) Use the x-values 0, 1, 2, 3, 4, 5, and 6 to create a table of x- and y-values.
   (b) Use Chart Wizard to create the corresponding graph.
   (c) Observing the result of part (b), state the value of the y-intercept. _____ In this example, the y-intercept is called the _____.
   (d) *Slope interpretation*. Studying the table and graph for parts (a) and (b), state what's happening to the y-values as the x-values increase by 1.
   (e) Explain the meaning of the answer to part (d) in terms of the *freezer's value*.

13. ***Simple depreciation: Linear decay***. An automobile costs $20,000.
   (a) If the automobile depreciates by $4000 per year, then write the equation that gives its value, y (in $thousands), after x years.
   (b) Use the x-values 0, 1, 2, 3, 4, and 5 to create a table of x- and y-values.
   (c) Use Chart Wizard to create the corresponding graph.
   (d) Observing the result of part (b), state the value of the y-intercept. _____ In this example, the y-intercept is called the _____.
   (e) *Slope interpretation*. Studying the table and graph for parts (b) and (c), state what's happening to the y-values as the x-values increase by 1.
   (f) Explain the meaning of the answer to part (e) in terms of the *automobile's value*.

14. ***Annual profits: Linear growth***. During its first year of operation, a company's profit was $8 million. Thereafter, the annual profit increased by $2 million per year.
   (a) If x denotes time (in years) with $x = 0$ corresponding to the first year, write the equation that gives the company's annual profit, y (in $millions), in terms of x.
   (b) Use the x-values 0, 1, 2, 3, 4, and 5 to create a table of x- and y-values.

(c) Use Chart Wizard to create the corresponding graph

(d) *Slope interpretation*. Studying the table and graph for parts (b) and (c), state what's happening to the y-values as the x-values increase by 1.

(e) Explain the meaning of the answer to part (d) in terms of *annual profits*.

1-2 Break-Even Analysis

Given the *cost function*, $C(x) = 2x + 3000$, and the *sales revenue function*, $R(x) = 5x$, where x denotes the number of units produced and sold, the *profit function* is determined by beginning with

$$P(x) = R(x) - C(x)$$

and substituting $5x$ for $R(x)$ and $2x + 3000$ for $C(x)$ to get

$$P(x) = 5x - (2x + 3000)$$

which simplifies to

$$P(x) = 3x - 3000.$$

Spreadsheet 1-2 gives tables of *x*- and *y*-values, along with a graph of each function.

SPREADSHEET 1-2

x	C=2x+3000	R=5x	P = R - C			
0	3000	0	-3000			
250	3500	1250	-2250			
500	4000	2500	-1500			
750	4500	3750	-750			
1000	5000	5000	0			
1250	5500	6250	750			
1500	6000	7500	1500			
1750	6500	8750	2250			
2000	7000	10000	3000			

Studying the contents of Spreadsheet 1-2, note that we chose not to add a legend in order to allow space for the graph. However, we can identify each line as follows: The revenue function passes through the origin, the cost function has a y-intercept of 3000, and the profit function is the lowest graph, having a y-intercept of -3000. Observe that the break-even point occurs at $x = 1000$ and that this is also the x-intercept of the profit function. Also, note that the y-intercept of the profit function is the negative fixed cost.

INSTRUCTIONS

Use the following instructions to create tables and graphs similar to those in Spreadsheet 1-2.

1. Type Labels
Label the columns of x-, C-, R-, and P-values as illustrated in Spreadsheet 1-2.

2. Create a Table of x- and y-values
After entering x-values, ***enter formulas*** for each function as illustrated in Spreadsheet 1-2. Then, ***copy the formulas*** as explained in the Excel Instructions of Section 1-1. Note that the formula for the profit function is either **=3*A2-3000** or **=C2-B2**, assuming that the column labels are typed in cells A1, B1, C1, and D1, the first x-value is typed in cell A2, the formula for the first C-value is typed in cell B2, the formula for the first R-value is typed in cell C2, and the formula for the first P-value is typed in cell D2. Refer to the earlier instructions if needed.

3. Create a Graph of the Data
Highlight the cells containing all the data and labels. Refer to the instructions in the previous section if needed.

EXERCISES

1. ***Break-even analysis.*** For a particular product, a firm has the following cost and sales revenue functions.

Cost function: $C(x) = 5x + 28$ Revenue function: $R(x) = 9x$

where x denotes the number of units produced and sold.
   (a) Verify that the ***profit function*** is given by $P(x) = 4x - 28$.
   (b) Use the x-values 0, 1, 2, 3, 4, 5, 6, 7, 8, 9, 10, 11, and 12 to create a table of x- and y-values for the cost, sales revenue, and profit functions as illustrated in Spreadsheet 1-2.
   (c) Explain why the y-values for the profit function can determined either by using the formula for the profit function or by entering a formula that subtracts cost function values from revenue function values.
   (d) Use Chart Wizard to create the corresponding graph.
   (e) Observing the result of part (d), identify the graph corresponding to each function.
   (f) Use algebra to show why the y-intercept of the profit function is the negative fixed cost.
   (g) Explain how the slope of a linear profit function can be determined from the cost and sales revenue functions.
   (h) Use algebra to verify that the x-coordinate of the break-even point is correct.

(i) Explain why the x-intercept of a profit function gives the x-coordinate of the break-even point.

2. **Break-even analysis.** For a particular product, a firm has the following cost and sales revenue functions.

   Cost function: $C(x) = 3x + 1500$ Revenue function: $R(x) = 8x$

where x denotes the number of units produced and sold.

(a) Verify that the *profit function* is given by $P(x) = 5x - 1500$.

(b) Use the x-values 0, 100, 200, 300, 400, 500, and 600 to create a table of x- and y-values for the cost, sales revenue, and profit functions.

(c) Use Chart Wizard to create the corresponding graphs.

(d) Observing the result of part (c), identify the graph corresponding to each function.

(e) Verify that the y-intercept of the profit function is the negative fixed cost and that the x-intercept of the profit function gives the x-coordinate of the break-even point.

3. **Decreasing the fixed cost by $1 increases profits by $1.** The following spreadsheet gives tables of x- and y-values for the profit functions $P(x) = 6x - 300$ and $P(x) = 6x - 299$, where x denotes the number of units produced and sold.

x	P=6x-300	P=6x-299				
0	-300	-299				
25	-150	-149				
50	0	1				
75	150	151				
100	300	301				
125	450	451				

(a) State the fixed cost for each profit function.

(b) Comparing the tables of y-values for both profit functions, explain why *decreasing the fixed cost by $1 results in a $1 increase in profit*. This is why, throughout the decade of the 1990s, many companies have downsized to decrease their fixed costs and increase profits.

4. **Decreasing the fixed cost by $1 increases profits by $1.** Consider the profit functions $P(x) = 10x - 400$ and $P(x) = 10x - 399$, where x denotes the number of units produced and sold.

(a) Use the x-values 0, 20, 40, 60, 80, 100, and 120 to create a table of x- and y-values for both profit functions.

(b) Use Chart Wizard to create the corresponding graphs.

(c) Observing the result of part (b), identify the graph corresponding to each function.

(d) State the fixed cost for each profit function.

(e) Comparing the tables of y-values for both profit functions, explain why *decreasing the fixed cost by $1 results in a $1 increase in profit*.

18

5. ***Decreasing the fixed cost results in a lower break-even point.*** Consider the profit functions $P(x) = 6x - 360$ and $P(x) = 6x - 240$, where x denotes the number of units produced and sold.

(a) Use the x-values 0, 20, 40, 60, 80, 100, and 120 to create a table of x- and y-values for both profit functions.
(b) Use Chart Wizard to create the corresponding graphs.
(c) Observing the result of part (b), identify the graph corresponding to each function.
(d) State the fixed cost for each profit function.
(e) State the x-coordinate of the break-even point for each function. Use algebra to verify that these values are correct.
(f) Comparing the x-coordinates of the break-even points for both profit functions, can we conclude that decreasing the fixed cost results in a lower break-even point?

6. ***Decreasing the fixed cost results in a lower break-even point.*** Consider the profit functions $P(x) = 5x - 400$ and $P(x) = 5x - 300$, where x denotes the number of units produced and sold.

(a) Use the x-values 0, 20, 40, 60, 80, 100, and 120 to create a table of x- and y-values for both profit functions.
(b) Use Chart Wizard to create the corresponding graphs.
(c) Observing the result of part (b), identify the graph corresponding to each function.
(d) State the fixed cost for each profit function.
(e) State the x-coordinate of the break-even point for each function. Use algebra to verify that these values are correct.
(f) Comparing the x-coordinates of the break-even points for both profit functions, can we conclude that decreasing the fixed cost results in a lower break-even point?

1-3 Using Excel's Goal Seek to:
(A) Solve for *x*, Given *y*; (B) Solve for *y*, Given *x*.

Suppose x and y are related by the equation

$$3x + 4y = 84$$

and we seek to determine the *x*-value corresponding to $y = 6$. Excel's *Goal Seek* is a tool used to determine an *input value* needed to produce a desired *output value*. Use the following instructions to solve this problem.

INSTRUCTIONS

Type the column labels **x** and **y** in cells A1 and B1, respectively. Enter the *y*-value **6** in cell B2 of the *y*-column as illustrated in Spreadsheet 1-3.

In cell C2, type the formula **=3*A2+4*B2** and press **Enter**, and the result, 24, appears in cell C2 as shown in Spreadsheet 1-3.

SPREADSHEET 1-3

x	y					
	6	24				

If not already there, use the mouse to *move the dark-bordered rectangle to the cell containing the formula*—in this case cell C2, which currently contains the value 24.

Select **Tools**, then choose *Goal Seek* and a dialog box appears. Note that the *cell containing the formula*—in this case cell C2, appears in the *Set cell* text box. Because it is our goal to *set this cell equal to a value of 84 by changing the value of cell A2*, we type **84** in the *To value* text box, type **A2** in the *By changing cell* text box, and click **OK**, and the required *x*-value (in this case 20) appears in cell A2.

This result is verified by substituting into the formula as follows:

$$3x + 4y = 84$$
$$3(20) + 4(6) = 84.$$

EXERCISES

For each of the following equations, use *Goal Seek* to
   (a) Find the *x*-value corresponding to the given *y*-value. Check your result by substituting into the equation.

(b) Find the y-value corresponding to the given x-value. Check your result by substituting into the equation.

1. $3x + 4y = 38$: (A) $y = 8$ (B) $x = 6$.

2. $2x + 5y = 36$: (A) $y = 4$ (B) $x = 3$.

3. $5x - 2y = 18$: (A) $y = 9$ (B) $x = 4$.

4. $6x - y = 16$: (A) $y = 8$ (B) $x = 3$.

Demand equation. Each of the following equations relates the unit price, p, of a product to the number of units demanded, q, of that product. For each of the following demand equations, use *Goal Seek* to

(a) Find the *unit price* corresponding to a demand of 6 units. Check your result by substituting into the equation.
(b) Find the *demand* corresponding to a unit price of $4. Check your result by substituting into the equation.

5. $2p + 3q = 60$ 6. $p + 5q = 40$

7. $3p + 4q = 60$ 8. $4p + 2q = 60$

Supply equation. Each of the following equations relates the unit price, p, of a product to the number of units supplied, q, of that product. For each of the following supply equations, use *Goal Seek* to

(a) Find the *unit price* corresponding to a supply of 2 units. Check your result by substituting into the equation.
(b) Find the *supply* corresponding to a unit price of $12. Check your result by substituting into the equation.

9. $4p - 3q = 12$ 10. $5p - 4q = 20$

1-4 Inserting a Trendline

Spreadsheet 1-4 gives a table of x- and y-values, along with a plot of the corresponding data points (x, y). The graph also illustrates a straight line fit to the set of data points. Because the individual data points are scattered about the straight line, such a graph is called a *scatterdiagram*. The straight line is fit so that it captures the relationship between the x- and y-values. There are many ways of fitting a straight line to a set of data points. Specifically, we could choose any two data points, connect them with a straight line, compute the slope of the line, and determine the equation of the straight line. By altering our choice of data points, we could fit many different straight lines to the same set of data points.

   A better way of fitting a straight line to a set of data points involves making an assessment of the *goodness of fit* of the straight line to the data points. This is done by considering, for each data point, the *vertical distance between the data point and the fitted line*. Such vertical distances are called **residuals** and indicate the extent to which the fitted line does <u>not</u> fit the data points. An overall measure of the extent to which a fitted line does <u>not</u> fit the data points is given by the sum of the squares of the residuals. Thus, the straight line that *minimizes the sum of the squares of the residuals* is the *best-fitting* straight line to the set of data points. Such a best-fitting straight line is appropriately called a *least-squares line*. Excel's *Trendline* fits a least-squares line to a set of data points. We hasten to add that *Trendline* is not limited to fitting straight lines to points but can also be used to fit quadratic and exponential functions, as well as others.

SPREADSHEET 1-4

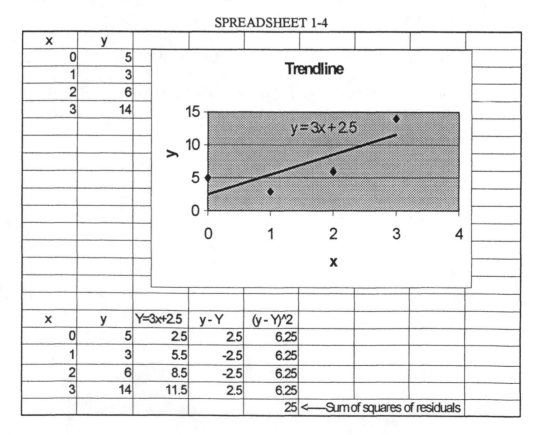

x	y
0	5
1	3
2	6
3	14

x	y	Y=3x+2.5	y - Y	(y - Y)^2		
0	5	2.5	2.5	6.25		
1	3	5.5	-2.5	6.25		
2	6	8.5	-2.5	6.25		
3	14	11.5	2.5	6.25		
				25	<—Sum of squares of residuals	

The bottom portion of Spreadsheet 1-4 gives the x- and y-values of the original data points, as well as the Y-coordinates of the fitted trend line. We use an upper-case Y for the Y-coordinates of the fitted trendline and a lower-case y for the y-values of the original data points. The Y-coordinates of the fitted trendline are determined by substituting the x-values into the equation of the fitted trendline. The residuals, $y - Y$, are computed by subtracting the Y-values of the fitted trendline from the y-values of the original data points. The last column, $(y - Y)^2$, gives the squares of the residuals and their corresponding sum. Fitting the trendline $Y = 3x + 2.5$ to this set of data points minimizes the sum of the squares of the residuals, and that minimum value is 25, as given at the bottom of the spreadsheet.

Using the Trendline

The trendline captures the relationship between the x- and y-values of the data points. The trendline's equation expresses that relationship. For example, if in Spreadsheet 1-4, x denotes a *test score* received by a salesperson on a sales aptitude test and y denotes the *first-year sales* (in $millions) for that salesperson, then the trendline's equation

$$y = 3x + 2.5$$

can be used to predict the first-year sales for a person who scores, say 1.5, by substituting 1.5 for x into the equation as follows:

$$y = 3(1.5) + 2.5$$
$$= 7 \quad \leftarrow \text{Predicted first-year sales (in \$millions) of a}$$
$$\text{salesperson with a test score of 1.5.}$$

INSTRUCTIONS

Use the following instructions to create tables and graphs similar to those in Spreadsheet 1-4.

<u>**1.** Create a Table of x- and y-values</u>

<u>**2.** Create a Graph of the Data</u>
After highlighting labels and data, select **Chart Wizard** from the toolbar, and a dialog box appears.

Step 1: Select **XY(Scatter)** in the Chart type section. Move to the Chart sub-type section and click on the first graph in the first column. Click **Next**.

Step 2: Click **Next**.

Step 3: Type the name you wish to give the graph in the Chart title box. We chose the name *trendline*. Label the x- and y-axes by entering **x** in the Value(X) axis section and **y** in the Value(Y) axis section. Delete the legend by clicking **Legend** at the top of the dialog box and then clicking the box next to the Show legend section to remove the checkmark from the box. Click **Next**.

Step 4: Click **Finish** and the graph should appear in your spreadsheet.

3. Change the Size of the Graph
Move the mouse pointer to the middle handle at the bottom of the chart until the pointer becomes a vertical double-sided arrow. Click, hold, and drag the line downward to enlarge the chart vertically. To enlarge the chart horizontally, move the mouse pointer to the middle handle at the side of the chart until the pointer becomes a horizontal double-sided arrow. Click, hold, and drag the line horizontally to enlarge the chart. Dragging the line horizontally in the reverse direction will decrease the size of the chart.

4. Insert a Trendline
4.1 Use the right mouse button to click on any data point inside the graph and select *Add Trendline*. Or select *Chart* from the Menu bar and then select *Add Trendline*.

4.2 Select the top left *linear* graph. Note that polynomial, exponential, or other graphs could be selected. Click *Options* and click the box next to *Display Equations on Chart* to place a checkmark in the box. This displays the equation on the chart. Click **OK**, and the trendline and its equation appear in the graph.

The columns in the bottom portion of Spreadsheet 1-4 are created by entering and copying formulas. The *sum* of the $(y - Y)^2$ column is computed by moving the dark-bordered rectangle to the cell that is to contain the sum, highlighting the cells containing the data in the column to be summed, and clicking $\sum$ in the Standard toolbar.

EXERCISES

1 – 3. For each of the following sets of data:
(a) Use Excel to create a graph, insert a trendline, and display its equation.
(b) Use the equation of the trendline to predict the y-value corresponding to $x = 3.5$.

1.

x	y
2	10
5	15
8	25
9	30

2.

x	y
3	9
1	5
2	7
5	14

3.

x	y
2	8
4	14
8	30
9	40

4. *Effect of advertising on sales (nonzero slope).* The table below gives monthly sales, y (in $thousands), corresponding to advertising expenditures, x (in $thousands).

Advertising expenditures x	0	1	2	3	4	5	6
Monthly sales y	3	9	9	15	17	25	27

(a) Use Excel to create a graph, fit a trendline, and display its equation for this set of data.
(b) Interpret the slope.

(c) Does the trendline indicate the existence of a relationship between advertising expenditures and monthly sales? In other words, do advertising expenditures appear to have any effect on monthly sales?

5. *Effect of advertising on sales (zero slope).* The table below gives monthly sales, y (in $thousands), corresponding to advertising expenditures, x (in $thousands).

Advertising expenditures x	0	1	2	3	4	5	6
Monthly sales y	4	6	2	4	2	6	4

(a) Use Excel to create a graph, fit a trendline, and display its equation for this set of data.
(b) Interpret the slope.
(c) Does the trendline indicate the existence of a relationship between advertising expenditures and monthly sales? In other words, do advertising expenditures appear to have any effect on monthly sales?

6. *Process control.* A company makes metal rods to length specifications of 3.00 cm $\pm$ 0.05 cm. This means that a metal rod is acceptable if its length lies within the interval 3.00 cm $\pm$ 0.05 cm. To continually monitor the production process, the company randomly selects a metal rod from the production process every half-hour and measures its length. The following table gives rod lengths, y, versus time, x (in half-hour increments).

Time x	0	1	2	3	4	5	6
Rod length y	3	3.01	2.99	3	2.99	3.01	3

(a) Use Excel to create a graph, fit a trendline, and display its equation for this set of data.
(b) Interpret the slope.
(c) Does the trendline indicate the existence of a relationship between rod length and time? In other words, does the passage of time appear to have any effect on rod length?

CHAPTER TWO

Power Functions and Quadratic Functions

2-1 Power Functions

We begin by studying tables of x- and y-values and graphs of functions defined by equations of the form

$$y = x^n$$

where n is a *positive integer* and $n \geq 2$. Specific examples include $y = x^2$, $y = x^3$, $y = x^4$, $y = x^5, \dots$. Because equations of this form give y as a *power* of x, they are appropriately called *power functions*. By restricting the power, n, to be a positive integer such that $n \geq 2$, we are, for the time being, excluding negative powers of x and fractional powers. We consider negative powers in a later section and fractional powers when appropriate. Also, it should be noted that equations of the form

$$y = ax^n$$

where a is a constant multiplier, also define power functions and will be included in this section. Specific examples include $y = 5x^2$, $y = 4x^3$, $y = -3x^5$, $y = -6x^8, \dots$. Finally, we note that equations such as $y = 7x$ and $y = -5x$ also define power functions. However, because the power is 1, these are linear functions and are not considered here. (They were discussed in Chapter 1.)

Positive Even Powers
We begin by discussing graphs of equations of the form

$$y = x^n$$

where n is a positive *even* integer. Examples include $y = x^2$, $y = x^4$, $y = x^6, \dots$. Spreadsheet 2-1 contains a table of x- and y-values of

$$y = x^2.$$

SPREADSHEET 2-1

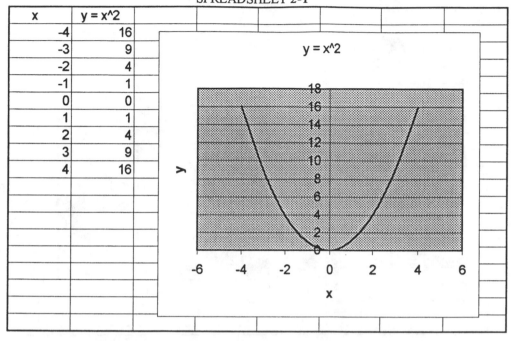

x	y = x^2					
-4	16					
-3	9					
-2	4					
-1	1					
0	0					
1	1					
2	4					
3	9					
4	16					

Studying the table and graph of Spreadsheet 2-1, note that negative x-values as well as positive x-values result in positive y-values. Of course, this is because a *negative number raised to an even power results in a positive number.* Notice that the y-value corresponding to $x = -4$ is the same as that corresponding to $x = 4$. The same holds for $x = -3$ and $x = 3$, $x = -2$ and $x = 2$, and $x = -1$ and $x = 1$. This is why the graph of $y = x^2$ is *symmetric*, the left branch being the mirror image of the right, and vice versa. We summarize by noting that graphs of

$$y = x^n,$$

where n is a *positive even power,* resemble the graph of $y = x^2$ given in Spreadsheet 2-1.

Positive Odd Powers

We begin by discussing graphs of equations of the form

$$y = x^n$$

where n is a positive *odd* integer. Examples include $y = x^3$, $y = x^7$, $y = x^9$, ...

27

Spreadsheet 2-2 contains a table of *x*- and *y*-values of

$$y = x^3.$$

SPREADSHEET 2-2

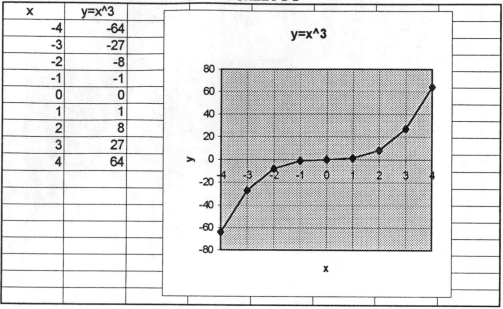

x	y=x^3
-4	-64
-3	-27
-2	-8
-1	-1
0	0
1	1
2	8
3	27
4	64

Studying the table and graph of Spreadsheet 2-2, note that negative *x*-values result in negative *y*-values, whereas positive *x*-values result in positive *y*-values. Of course, this is because a *negative number raised to an odd power results in a negative number, whereas a positive number raised to an odd power results in a positive number.* Notice that the *y*-value corresponding to *x* = -4 is the negative of that corresponding to *x* = 4. The same holds for *x* = -3 and *x* = 3, *x* = -2 and *x* = 2, and *x* = -1 and *x* = 1. We summarize by noting that graphs of

$$y = x^n,$$

where *n* is a *positive odd power,* resemble the graph of $y = x^3$ given in Spreadsheet 2-2.

Finally, as mentioned earlier, we consider equations of the form

$$y = ax^n,$$

where *a* is a constant multiplier. To illustrate the effect of the constant multiplier *a* upon the graph, Spreadsheet 2-3 contains tables of *x*- and *y*-values and graphs of $y = x^2$ versus $y = 3x^2$. Observe that these graphs are similar except that the graph of $y = 3x^2$ is

28

thinner than that of $y = x^2$. This is because the *y*-values of $y = 3x^2$ are 3 *times* those of $y = x^2$.

SPREADSHEET 2-3

x	y=x^2	y=3x^2
-4	16	48
-3	9	27
-2	4	12
-1	1	3
0	0	0
1	1	3
2	4	12
3	9	27
4	16	48

y=x^2 vs y=3x^2

INSTRUCTIONS

Use the following instructions to create tables and graphs similar to those in Spreadsheet 2-1.

1. Type Labels
1.1 As an example, we will create a table of *x*- and *y*-values for the function featured in Spreadsheet 2-1. Thus, we will use Column A for the *x*-values and Column B for the *y*-values of the equation $y = x^2$.

1.2 Once you have a blank worksheet, use the mouse to move your pointer to cell A1 and click the left mouse button to make that cell the active cell. A rectangle with a dark border should appear around cell A1. Use the spacebar to move to the middle of the cell and type **x**. This labels Column A. Now, label Column B with its equation by moving the dark-bordered rectangle around cell B1 and typing **y=x^2** in cell B1.

2. Create a Table of *x*- and *y*-values
2.1 Beginning with cell A2, type the *x*-values **-4** through **4** into Column A as illustrated in Spreadsheet 2-1.

2.2 Now, we *enter a formula* to compute the corresponding *y*-values for the equation $y = x^2$ in Column B. Move the dark-bordered rectangle to cell B2 and type the formula **=A2^2**. Then press **Enter**. Note that the symbol ^ means exponentiation. Although it is

29

not yet needed, the symbol * means multiplication. The formula =A2^2 entered in cell B2 returns to cell B2 the y-value ($y = 16$) corresponding to the x-value ($x = -4$) of cell A2.

2.3 Now, we *copy the formula* down through cell B10. Move the dark-bordered rectangle to cell B2. Use the mouse to move the pointer to the small black box (called a handle) at the lower right corner of cell B2. The mouse pointer becomes a thick black plus. Click the mouse button without releasing it and drag the mouse pointer down to cell B10. Release the mouse button at cell B10, and cells B3 to B10 will contain the formula values.

3. Create a Graph of the Data

3.1 Move the mouse pointer to the middle of cell A1, click, hold, and drag the mouse pointer until it highlights the cells containing the data and labels. Cell A1 will remain unhighlighted.

3.2 Select **Chart Wizard** from the toolbar, and a dialog box appears.

Step 1: Select **XY(Scatter)** in the Chart type section. Move to the Chart sub-type section and click on the first graph in the second column. Click **Next**.

Step 2: Click **Next**.

Step 3: Enter **x** in the Value(X) axis section and **y** in the Value(Y) axis section to label the x- and y-axes. Click **Next**.

Step 4: Click **Finish,** and the graph should appear in your spreadsheet.

4. Change the Size of the Graph

Move the mouse pointer to the middle handle at the bottom of the chart until the pointer becomes a vertical double-sided arrow. Click, hold, and drag the line downward to enlarge the chart vertically. To enlarge the chart horizontally, move the mouse pointer to the middle handle at the side of the chart until the pointer becomes a horizontal double-sided arrow. Click, hold, and drag the line horizontally to enlarge the chart. Dragging the line horizontally in the reverse direction will decrease the size of the chart.

EXERCISES

1. *Power functions: Even powers.* Create a table of x- and y-values for the function $y = x^4$.
   - (a) Use the x-values -4, -3, -2, -1, 0, 1, 2, 3, 4.
   - (b) Use Chart Wizard to create the corresponding graph.
   - (c) Is the graph similar in appearance to that of $y = x^2$ in Spreadsheet 2-1?

2. *Power functions: Even powers.* Create a table of x- and y-values for the function $f(x) = x^6$.
   - (a) Use the x-values -4, -3, -2, -1, 0, 1, 2, 3, 4.
   - (b) Use Chart Wizard to create the corresponding graph.

(c) Is the graph similar in appearance to that of $y = x^2$ in Spreadsheet 2-1?

3. <u>Pencil and Paper Exercise</u>. Draw a graph that typifies graphs of $f(x) = x^n$, where n is a *positive even integer*.

4. ***Power functions: Odd powers.*** Create a table of x- and y-values for the function $y = x^5$.
   (a) Use the x-values -4, -3, -2, -1, 0, 1, 2, 3, 4.
   (b) Use Chart Wizard to create the corresponding graph.
   (c) Is the graph similar in appearance to that of $y = x^3$ in Spreadsheet 2-2?

5. <u>Pencil and Paper Exercise</u>. Draw a graph that typifies graphs of $f(x) = x^n$, where n is a *positive odd integer* such that $n \geq 3$.

6. ***Effect of constant multiplier.*** Create a table of x- and y-values for the functions $y = x^2$ and $y = 4x^2$.
   (a) Use the x-values -4, -3, -2, -1, 0, 1, 2, 3, 4.
   (b) Use Chart Wizard to create the corresponding graphs. Remember to highlight both columns of y-values—one for x^2 and the other for 4*x^2—in addition to the column containing the x-values.
   (c) Observe the table to determine the y-intercepts of both equations and verify this result by checking the graphs.
   (d) State which equation has the thinner graph. Multiplying x^2 by 4 has what effect on the graph of $y = x^2$?

7. ***Effect of constant multiplier.*** Create a table of x- and y-values for the functions $y = x^2$ and $y = \frac{1}{2}x^2$.
   (a) Use the x-values -4, -3, -2, -1, 0, 1, 2, 3, 4.
   (b) Use Chart Wizard to create the corresponding graphs. Remember to highlight both columns of y-values—one for x^2 and the other for 0.5*x^2—in addition to the column containing the x-values.
   (c) Observe the table to determine the y-intercepts of both equations and verify this result by checking the graphs.
   (d) State which equation has the thinner graph. Multiplying x^2 by 1/2 has what effect on the graph of $y = x^2$?

8. <u>Pencil and Paper Exercise</u>. Draw the graphs of $f(x) = x^2$ and $f(x) = 5x^2$. Label each graph with its equation.

9. <u>Pencil and Paper Exercise</u>. Draw the graphs of $f(x) = x^2$ and $f(x) = \frac{1}{4}x^2$. Label each graph with its equation.

10. *Comparing even powers.* Create a table of x- and y-values for the functions $y = x^2$ and $y = x^4$.

  (a) Use the x-values -1.2, -1.1, -1, -0.9, . . . , 1.2.
  (b) Use Chart Wizard to create the corresponding graphs.
  (c) Observe the table of y-values for both equations and state the equation of the lower graph on the interval $-1 < x < 1$. Explain why its equation results in the lower graph on this interval.

11. *Comparing even powers.* Repeat Exercise 10 for the functions $y = x^2$ and $y = x^6$.

12. *Experimenting with graphing options.* Create a table of x- and y-values for the functions $y = x^2$ and $y = x^3$.

  (a) Use the x-values 0, 0.1, 0.2, 0.3, 0.4, . . . , 1.
  (b) Use Chart Wizard to create the corresponding graphs.

> While in Step 1 of <u>Creating a Graph of the Data</u>, after selecting **XY(Scatter)** in the Chart type section, move to the Chart sub-type section and click on the different graphs. For example, click on the second graph in the first column and note how it differs from the current graph. Note the description of each graph type in the box below.

> While in Step 3 of <u>Creating a Graph of the Data</u>, click on Gridlines and experiment with various options. Note how gridlines can be removed or increased.

> While in Step 3 of <u>Creating a Graph of the Data</u>, click on Legend and note how the legend can be included or removed.

> While in Step 3 of <u>Creating a Graph of the Data</u>, click in the title box to add a title to the graph.

  (c) Observe the table of y-values for both equations and state the equation of the lower graph on the interval $0 < x < 1$.

13. *Experimenting with graphing options.* Repeat Exercise 12 for the same two functions on the interval $-1 < x < 0$ using the x-values -1, -0.9, -0.8, . . ., -0.1, 0.

2-2 Graphing Concepts: Shifts and Reflections

Here we discuss graphing concepts that enable us to sketch graphs of functions whose equations are derived from equations with known graphs.

Vertical Shifts
Spreadsheet 2-4 contains a table of x- and y-values, along with corresponding graphs of $y = x^2$ and $y = x^2 + 5$.

SPREADSHEET 2-4

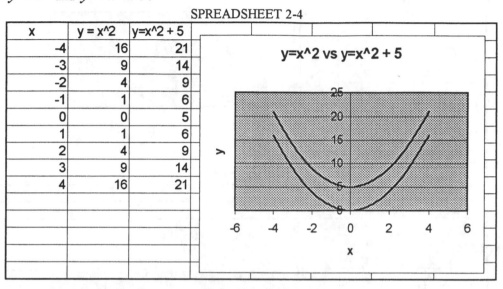

x	y = x^2	y=x^2 + 5
-4	16	21
-3	9	14
-2	4	9
-1	1	6
0	0	5
1	1	6
2	4	9
3	9	14
4	16	21

Observe in Spreadsheet 2-4 that the graph of $y = x^2 + 5$ is 5 units higher than the graph of $y = x^2$. This is because 5 is added to the equation $y = x^2$ to obtain $y = x^2 + 5$, and therefore, as revealed in the table of Spreadsheet 2-4, the y-values of $y = x^2 + 5$ are 5 greater than those of $y = x^2$. Thus, the graph of $y = x^2 + 5$ is obtained from that of $y = x^2$ by *lifting the graph of* $y = x^2$ *vertically by 5 units*. This is called a ***vertical shift***. Analagously, we will discover in the homework exercises that the y-values of $y = x^2 - 5$ are 5 less than those of $y = x^2$, and therefore, the graph of $y = x^2 - 5$ is obtained from that of $y = x^2$ by *lowering the graph of* $y = x^2$ *vertically by 5 units.*

Horizontal Shifts
Let's begin with $y = x^2$ and replace x with $x - 3$ to obtain $y = (x - 3)^2$. We are concerned with how the graph of $y = (x - 3)^2$ is drawn from that of $y = x^2$.
Spreadsheet 2-5 contains tables of x- and y-values and graphs of both functions. Observe that both graphs are the same except for the fact that the graph of $y = (x - 3)^2$ is

33

centered at $x = 3$, whereas the graph of $y = x^2$ is centered at $x = 0$. Thus, replacing x with $x - 3$ *shifts the graph horizontally to the right by 3 units.*

SPREADSHEET 2-5

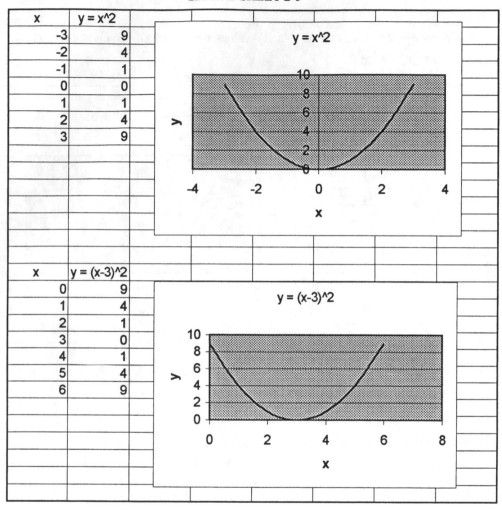

x	y = x^2
-3	9
-2	4
-1	1
0	0
1	1
2	4
3	9

x	y = (x-3)^2
0	9
1	4
2	1
3	0
4	1
5	4
6	9

Reflections in the x-axis

Spreadsheet 2-6 contains tables of x- and y-values and graphs of $y = x^2$ and $y = -x^2$.

Note that the y-values of $y = -x^2$ are negatives of those of $y = x^2$. Thus, to note the effect of a *negative constant multiplier* on the graph of a function, observe that multiplying x^2 by -1 has *turned the graph of* $y = x^2$ *upside down*. This is called a *reflection in the x-axis*.

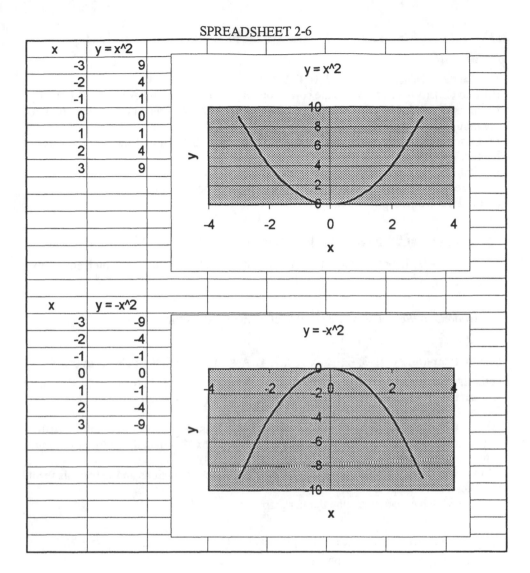

x	y = x^2
-3	9
-2	4
-1	1
0	0
1	1
2	4
3	9

x	y = -x^2
-3	-9
-2	-4
-1	-1
0	0
1	-1
2	-4
3	-9

EXERCISES

1. *Vertical shift.* Create a table of x- and y-values for the functions $y = x^2$ and $y = x^2 + 7$.

(a) Use the x-values -4, -3, -2, -1, 0, 1, 2, 3, 4.

(b) Use Chart Wizard to create the corresponding graphs. Remember to highlight both columns of y-values—one for x^2 and the other for x^2 + 7—in addition to the column containing the x-values.

(c) Observe the table to determine the y-intercepts of both equations and verify this result by checking the graphs.

(d) State the equation of the higher graph.

(e) State the relationship between the y-coordinates of these graphs.

35

2. *Vertical shift.* Create a table of x- and y-values for the functions $y = x^2$ and
$y = x^2 - 4$.

(a) Use the x-values -4, -3, -2, -1, 0, 1, 2, 3, 4.
(b) Use Chart Wizard to create the corresponding graphs. Remember to highlight both columns of y-values—one for x^2 and the other for x^2 – 4—in addition to the column containing the x-values.
(c) Observe the table to determine the y-intercepts of both equations and verify this result by checking the graphs.
(d) State the equation of the higher graph.
(e) State the relationship between the y-coordinates of both graphs.
(f) Find the x-intercept(s) of each graph.

3. Pencil and Paper Exercise. Draw the graphs of $f(x) = x^2$, $f(x) = x^2 + 9$, and $f(x) = x^2 - 9$. Label each graph with its equation. Label any x- and y-intercepts with their coordinates.

4. *Reflections in the x-axis.* Create a table of x- and y-values for $y = -x^2$ and $y = -4x^2$.

(a) Use the x-values -4, -3, -2, -1, 0, 1, 2, 3, 4. Assuming the first x-value is in cell A2, enter the formula -1*A2^2 in cell B2 to ensure that x^2 is preceded by a negative sign. Experiment by entering -A2^2 and note the error (the negative sign is missing). Remember this for future formulas.
(b) Use Chart Wizard to create the corresponding graphs. Remember to highlight both columns of y-values—one for -x^2 and the other for -4*x^2—in addition to the column containing the x-values.
(c) Observe the table to determine the y-intercepts of both equations and verify this result by checking the graphs.
(d) State which equation has the thinner graph.

5. Pencil and Paper Exercise. Draw the graphs of $f(x) = -x^2$ and $f(x) = -5x^2$. Label each graph with its equation.

6. *Horizontal shift.* Create a table of x- and y-values for the functions $y = x^2$ and $y = (x+3)^2$

(a) Use the x-values -6, -5, -4, -3, -2, -1, 0. Create separate columns of x- and y-values for each function as illustrated in Spreadsheet 2-5.
(b) Use Chart Wizard to create a separate graph for each function.
(c) State the x-intercept of each graph. Replacing x with $x + 3$ has what effect on the graph of $y = x^2$?

7. <u>Pencil and Paper Exercise</u>. Draw the graphs of $f(x) = x^3$, $f(x) = (x+4)^3$, and $f(x) = (x-4)^3$. Label each graph with its equation. Label any x- and y-intercepts with their coordinates.

8. <u>Pencil and Paper Exercise</u>. Draw the graphs of $y = x^2$, $y = (x-5)^2$, and $y = (x+5)^2$. Label each graph with its equation. Label any x- and y-intercepts with their coordinates.

9. <u>Pencil and Paper Exercise</u>. Draw the graphs of $y = x^2$, $y = x^2 + 8$, and $y = x^2 - 8$.
Label each graph with its equation. Label any x- and y-intercepts with their coordinates.

10. <u>Pencil and Paper Exercise</u>. Draw the graphs of $y = -x^2$, $y = -x^2 + 8$, and $y = -x^2 - 8$. Label each graph with its equation. Label any x- and y-intercepts with their coordinates.

2-3 Quadratic Functions: Graphs of $y = ax^2 + bx$ and $y = ax^2 + bx + c$

A *quadratic function* is defined by an equation of the form

$$y = ax^2 + bx + c$$

where a, b, and c are constants such that $a \neq 0$. The following are examples of quadratic functions:

$$y = -x^2 + 5x - 6 \quad \text{where } a = -1, \ b = 5, \text{ and } c = -6.$$

$$y = x^2 - 3x \quad \text{where } a = 1, \ b = -3, \text{ and } c = 0.$$

$$y = -3x^2 + 4 \quad \text{where } a = -3, \ b = 0, \text{ and } c = 4.$$

Graphs of $y = ax^2 + bx$

Spreadsheet 2-7 gives a table of x- and y-values, along with a graph of $y = x^2 - 6x$, which can be written in factored form as $y = x(x - 6)$.

SPREADSHEET 2-7

x	y=x(x - 6)
-1	7
0	0
1	-5
2	-8
3	-9
4	-8
5	-5
6	0
7	7

y=x^2-6x or y=x(x-6)

Observing the graph and table of x- and y-values, note that the graph passes through the origin, which is one of the x-intercepts. The other x-intercept occurs at $x = 6$. Because of symmetry, the x-coordinate of the *vertex* is *midway between both x-intercepts*, at $x = 3$.

Also, observe that the y-coordinate of the vertex is -9. It's good practice to confirm these results algebraically, as we will be asked to do in the homework exercises. Additionally, we will be asked to generalize certain results for the form $y = ax^2 + bx$.

Comparing Graphs of $y = ax^2 + bx$ ***and*** $y = ax^2 + bx + c$

Spreadsheet 2-8 gives a table of x- and y-values, along with graphs of $y = x^2 - 10x$ and $y = x^2 - 10x + 30$. It will be helpful to factor out x in each equation so that

$$y = x^2 - 10x \qquad \text{is rewritten as} \qquad y = x(x - 10)$$

and

$$y = x^2 - 10x + 30 \qquad \text{is rewritten as} \qquad y = x(x - 10) + 30.$$

SPREADSHEET 2-8

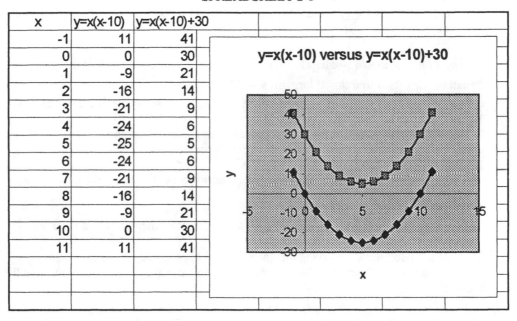

x	y=x(x-10)	y=x(x-10)+30
-1	11	41
0	0	30
1	-9	21
2	-16	14
3	-21	9
4	-24	6
5	-25	5
6	-24	6
7	-21	9
8	-16	14
9	-9	21
10	0	30
11	11	41

Observing the graphs and table of x- and y-values in Spreadsheet 2-8, note that the graph of $y = x^2 - 10x + 30$ can be obtained from that of $y = x^2 - 10x$ by *shifting* the graph of $y = x^2 - 10x$ *upward* by 30 units. Because of the vertical shift, the ***x-coordinate*** of the ***vertex*** point is the ***same*** for both graphs. Of course, the y-coordinates of these graphs differ by 30, the amount of the vertical shift.

39

EXERCISES

1. Create a table of x- and y-values for the function $y = -x^2 + 8x$. Note that the right-hand side can be expressed in factored form as $y = x(-x + 8)$.

(a) Use the x-values 0, 1, 2, 3, 4, 5, 6, 7, 8. Assuming the first x-value is located in cell A2, one formula for the corresponding y-value is =A2*(-A2+8), which is typed in cell B2. Give another formula for the y-value.

(b) Use Chart Wizard to create the corresponding graph. Note that the parabola *opens down,* which is confirmed by the *negative* coefficient (a = -1) of x^2.

(c) Observing the graph, state the x-intercepts.

(d) <u>Pencil and Paper Exercise</u>. Confirm the answer to part (d) by using the factored form of the equation to find the x-intercepts.

(e) Observing the graph, state the x-coordinate of the vertex.

(f) <u>Pencil and Paper Exercise</u>. State how the x-coordinate of the vertex is related to the x-intercepts.

(g) State the y-intercept, observing the graph and the appropriate y-value.

(h) <u>Pencil and Paper Exercise</u>. Confirm the answer to part (g) by using algebra to find the y-intercept.

2. Create a table of x- and y-values for the function $y = x^2 - 8x$. Note that the right-hand side can be expressed in factored form as $y = x(x - 8)$.

(a) Use the x-values 0, 1, 2, 3, 4, 5, 6, 7, 8. Assuming the first x-value is located in cell A2, give two formulas for the corresponding y-value in cell B2.

(b) Use Chart Wizard to create the corresponding graph. State whether the parabola *opens up* or *down* and state why the answer makes sense.

(c) Observing the graph, state the x-intercepts.

(d) <u>Pencil and Paper Exercise</u>. Confirm the answer to part (d) by using the factored form of the equation to find the x-intercepts.

(e) Observing the graph, state the x-coordinate of the vertex.

(f) <u>Pencil and Paper Exercise</u>. State how the x-coordinate of the vertex relates to the x-intercepts.

(g) State the y-intercept, observing the graph and the appropriate y-value.

(h) <u>Pencil and Paper Exercise</u>. Confirm the answer to part (g) by using algebra to find the y-intercept.

3. Create a table of x- and y-values for the function $y = -2x^2 + 10x$. Note that the right-hand side can be expressed in factored form as $y = 2x(-x + 5)$.

(a) Use the x-values 0, 1, 2, 3, 4, 5. Assuming the first x-value is located in cell A2, give two formulas for the corresponding y-value in cell B2.

(b) Use Chart Wizard to create the corresponding graph. State whether the parabola *opens up* or *down* and state why the answer makes sense.

(c) Observing the graph, state the x-intercepts.

(d) <u>Pencil and Paper Exercise</u>. Confirm the answer to part (d) by using the factored form of the equation to find the x-intercepts.

(e) Observing the graph, state the x-coordinate of the vertex.

(f) Pencil and Paper Exercise. State how the x-coordinate of the vertex is related to the x-intercepts.

(g) State the y-intercept, observing the graph and the appropriate y-value.

(h) Pencil and Paper Exercise. Confirm the answer to part (g) by using algebra to find the y-intercept.

4. Pencil and Paper Exercise. For equations of the form $y = ax^2 + bx$:

(a) Use algebra to show why graphs of such equations always pass through the origin. In other words, one of the x-intercepts is $x = 0$.

(b) Use algebra to show why the other x-intercept is $x = \dfrac{-b}{a}$.

(c) Explain why the x-coordinate of the vertex is given by $\dfrac{-b}{2a}$.

(d) Explain why the y-intercept is $y = 0$.

5 – 8. Pencil and Paper Exercises. For each equation, draw the graph by using the following procedure:

Step 1. Determine whether the parabola opens up or down by noting the sign of a, the coefficient of x^2 . Remember, if a is *positive*, the parabola *opens up*; if a is *negative*, the parabola *opens down*.

Step 2. Find the x-intercepts.

Step 3. Find the x-coordinate of the vertex by determining the midpoint of the x-intercepts. Confirm the result by using the formula $\dfrac{-b}{2a}$. Substitute the x-coordinate into the quadratic equation to obtain the y-coordinate of the vertex.

Step 4. Draw the graph.

5. $y = x^2 - 4x$ 6. $y = -4x^2 + 32x$ 7. $y = x^2 + 10x$ 8. $y = -x^2 + 14x$

9. Create a table of x- and y-values for the functions $y = -x^2 + 4x$ and $y = -x^2 + 4x + 12$.

(a) Use the x-values -2, -1, 0, 1, 2, 3, 4, 5. Assuming the first x-value is located in cell A2, one formula for the corresponding y-value of $y = -x^2 + 4x$ is =A2*(-A2+4), which is typed in cell B2. Explain why a formula for the corresponding y-value of $y = -x^2 + 4x + 12$ is =B2+12, which is typed in cell C2.

(b) Explain why the *x-coordinate* of the *vertex* is the *same* for both graphs.

(c) The y-coordinates of these graphs differ by what number?

10. Create a table of x- and y-values for the functions $y = 2x^2 - 16x$ and

$y = 2x^2 - 16x + 40$.
  (a) Use the x-values 0, 1, 2, 3, 4, 5, 6, 7, 8.
  (b) Explain why the x-*coordinate* of the *vertex* is the *same* for both graphs.
  (c) The y-coordinates of these graphs differ by what number?

11. Create a table of x- and y-values for the functions $y = x^2 - 6x$ and

$y = x^2 - 6x + 9$.
  (a) Use the x-values -2, -1, 0, 1, 2, 3, 4, 5, 6, 7, 8.
  (b) Explain why the x-*coordinate* of the *vertex* is the *same* for both graphs.
  (c) The y-coordinates of these graphs differ by what number?

12. Create a table of x- and y-values for the functions $y = x^2 - 6x$ and

$y = x^2 - 6x + 11$.
  (a) Use the x-values -2, -1, 0, 1, 2, 3, 4, 5, 6, 7, 8.
  (b) Explain why the x-*coordinate* of the *vertex* is the *same* for both graphs.
  (c) The y-coordinates of these graphs differ by what number?

13 – 18. <u>Pencil and Paper Exercises</u>. Each equation that follows is of the form

$y = ax^2 + bx + c$. Draw the graph of each equation by using the following procedure:

Step 1. Determine whether the parabola opens up or down by noting the sign of a, the coefficient of x^2. Remember, if a is *positive*, the parabola *opens up*; if a is *negative*, the parabola *opens down*.

Step 2. Find the x-intercepts of the corresponding form $y = ax^2 + bx$.

Step 3. Find the x-coordinate of the vertex by determining the midpoint of the x-intercepts you found in part (b). Confirm the result by using the formula $\dfrac{-b}{2a}$. Substitute the x-coordinate into the quadratic equation of the form $y = ax^2 + bx + c$ to obtain the y-coordinate of the vertex.

Step 4. Draw the graph. Find the x-intercepts of the form $y = ax^2 + bx + c$, if they exist.

13. $y = x^2 - 8x + 16$ 14. $y = x^2 - 8x + 19$ 15. $y = x^2 + 8x + 16$

16. $y = x^2 + 8x + 19$ 17. $y = x^2 + 6x + 9$ 18. $y = x^2 + 6x + 4$

2-4 Revenue Functions and Break-Even Analysis

Sales revenue is determined by the formula

$$\textbf{\textit{Sales revenue = (unit price)(number of units sold)}}.$$

If $R(x)$ denotes total sales revenue gained from selling x units and p denotes unit price, then the above equation becomes

$$R(x) = px$$

If, for example, the unit price for some product is $5 (i.e., $p = 5$), then the sales revenue function is given by the linear equation

$$R(x) = 5x.$$

However, sometimes the unit price is not constant but, instead, depends upon the demand level for the product. In such situations, the relationship between unit price, p, and demand, x, is expressed by a demand equation. If, for example, we consider the demand equation

$$p = 10 - x,$$

then the sales revenue function becomes

$$\begin{aligned} R(x) &= px \\ &= (10 - x)x \\ &= 10x - x^2 \\ &= -x^2 + 10x. \end{aligned}$$

Spreadsheet 2-9 gives unit price, p, and sales revenue values, R, for given demand levels, x.

SPREADSHEET 2-9

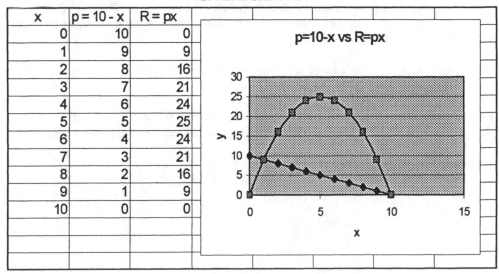

x	p = 10 - x	R = px
0	10	0
1	9	9
2	8	16
3	7	21
4	6	24
5	5	25
6	4	24
7	3	21
8	2	16
9	1	9
10	0	0

Observing the tables and graphs in Spreadsheet 2-9, note that the maximum sales revenue is $25 and it occurs (at $x = 5$) at a sales level of 5 units. Also, note that sales revenue decreases for $x > 5$. At first thought, this appears unreasonable until one observes (by scanning the p-values in the tables) that the unit price, p, decreases as the number of units sold, x, increases. This is illustrated by the straight line (demand equation) in the graph of Spreadsheet 2-9. In other words, the reason why sales revenue decreases is that as more and more units are being sold, they are being sold at a cheaper an cheaper unit price so that eventually (for $x > 5$), the sales revenue decreases despite the increase in sales volume.

Break-even Analysis
The top portion of Spreadsheet 2-10 gives tables and graphs for the Sales revenue function $R(x) = -x^2 + 10x$, along with the cost function $C(x) = x + 14$. Observe the intersection points of the sales revenue and cost functions. These points where sales revenue equals cost are appropriately called the **break-even points**. Scanning the R- and C-values in the tables, note that at $x = 2$, $R = C = 16$ and that at $x = 7$, $R = C = 21$. Thus, the break-even points occur at $x = 2$ and $x = 7$, so sales revenue equals cost at these x-values. The break-even points are verified algebraically by setting $R(x) = C(x)$ and solving for x.

Profit Function
By definition,

$$\text{Profit} = \text{revenue - cost}$$

or,

$$P(x) = R(x) - C(x).$$

The bottom portion of Spreadsheet 2-10 gives a table and graph for the profit function. The profits (P-values) are determined by subtracting the revenue and cost values corresponding to the x-values 2 through 7 given in the top portion of the spreadsheet. Observe that profit equals 0 at the break-even points $x = 2$ and $x = 7$. These are the x-intercepts of the profit function. Also, note that profit is positive for the x-values between the break-even points, $2 < x < 7$.

This is verified algebraically by determining the equation of the profit function and finding the x-intercepts by setting $P = 0$ and solving for x.

Finally, note that the *maximum profit* occurs at the vertex of the profit function. Of course, the x-coordinate of the vertex lies midway between the x-intercepts or, as we discussed in Section 2-3, it can be determined by using the formula $x = \dfrac{-b}{2a}$.

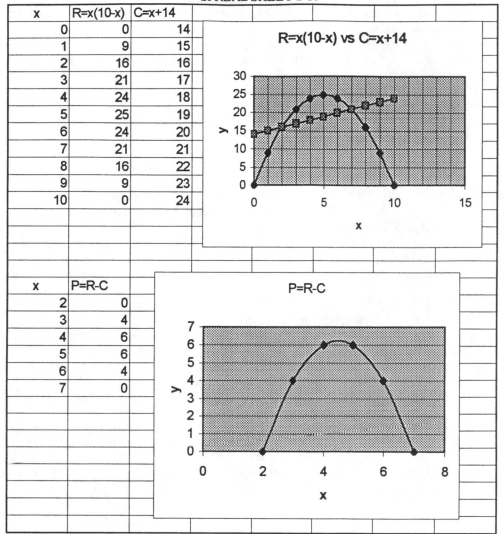

x	R=x(10-x)	C=x+14
0	0	14
1	9	15
2	16	16
3	21	17
4	24	18
5	25	19
6	24	20
7	21	21
8	16	22
9	9	23
10	0	24

x	P=R-C
2	0
3	4
4	6
5	6
6	4
7	0

EXERCISES

1. *Sales revenue function.* Consider the demand equation $p = 40 - 2x$, where p and x denote unit price and demand, respectively.

(a) Use the x-values 0, 1, 2, . . . , 20 to create a table of x-, p-, and R-values, where R denotes sales revenue. Assuming that cells A2, B2, and C2 will contain the first x-, p-, and R-values, respectively, type the formula **=40 – 2*A2** in cell B2 and the formula **=A2*B2** in cell C2.

(b) Use Chart Wizard to create the corresponding graphs.

(c) <u>Pencil and Paper Exercise</u>. Determine the equation of the sales revenue function, $R(x)$. Give another formula that could be typed in cell C2 in part (a).

45

(d) <u>Pencil and Paper Exercise</u>. Use the table created in the spreadsheet to determine the sales revenue gained from selling 4 units. Locate this point on the graph. State the unit price corresponding to this point.

(e) <u>Pencil and Paper Exercise</u>. Determine the maximum sales revenue and the number of units that should be sold in order to achieve the maximum sales revenue. Locate this point on the graph. State the unit price corresponding to this point.

(f) <u>Pencil and Paper Exercise</u>. Scan the column of unit prices, p. Does the unit price increase or decrease as the number of units sold increases?

(g) <u>Pencil and Paper Exercise</u>. Explain why the sales revenue decreases as the number of units sold increases on the interval $x > 10$.

2. ***Break-even analysis.*** Consider the cost and sales revenue functions $C(x) = 9x + 90$ and $R(x) = -3x^2 + 48x$, where x denotes number of units produced and sold.

(a) Use the x-values $0, 1, 2, \ldots, 16$ to create a table of x-, C-, and R-values.

(b) Use Chart Wizard to create the corresponding graphs.

(c) <u>Pencil and Paper Exercise</u>. Observe the table of x-, C-, and R-values to determine the break-even points. State the x-coordinates of the break-even points. Give the sales revenue and cost associated with each break-even point. Explain what it means to break even.

(d) <u>Pencil and Paper Exercise</u>. Verify the break-even points determined in part (c) by setting $R(x) = C(x)$ and solving for x.

(e) Using the table created in part (a), create a column of profit values (P-values) and use Chart Wizard to create the graph of the profit function. Assuming that cells A2, B2, and C2 will contain the first x-, C-, and R-values, respectively, type the formula =C2-B2 in cell D2 to create the column of profit function values in column D.

(f) <u>Pencil and Paper Exercise</u>. Determine the equation of the profit function, $P(x)$. Give another formula that could be typed in cell D2 in part (d).

(g) <u>Pencil and Paper Exercise</u>. Using the equation of the profit function, $P(x)$, determine the break-even points algebraically. Verify that they equal those in the spreadsheet and graph.

(h) <u>Pencil and Paper Exercise</u>. Study the equation of the profit function, $P(x)$. Explain or show algebraically why the *y-intercept* of the profit equation is the *negative fixed cost*.

3. ***Effect of a decrease in fixed cost on profit.*** Consider the profit functions $P(x) = -x^2 + 20x - 75$ and $P(x) = -x^2 + 20x - 65$, where $P(x)$ denotes the profit gained from producing x units.

(a) Use the x-values $0, 1, 2, \ldots, 20$ to create columns of y-values for both profit functions.

(b) Use Chart Wizard to create the corresponding graphs.

(c) <u>Pencil and Paper Exercise</u>. State the fixed cost for each profit equation.

(d) <u>Pencil and Paper Exercise</u>. Observing the tables of y-values and graphs for both profit functions, state the effect of decreasing the fixed cost by $10.

4. <u>Pencil and Paper Exercise</u>. Graph both profit functions $P(x) = -x^2 + 14x - 40$ and $P(x) = -x^2 + 14x - 35$ on the same set of axes. State the maximum profit and fixed cost for each. State the effect on profit of decreasing the fixed cost by $5.

CHAPTER 3

Selected Rational Functions

3-1 Odd and Even Powers of $y = \dfrac{a}{x^n}$; Sums of Functions

In this section, we will study graphs of functions of the form $y = \dfrac{a}{x^n}$, where a is a constant and n is a positive integer such that $n \geq 1$. We consider two cases—*odd and even powers*. Because these functions are defined as quotients of two polynomials, they belong to a category of functions called rational functions.

Odd Powers

The graph of $y = \dfrac{1}{x}$ typifies the graphs of $y = \dfrac{a}{x^n}$, where a is positive and n is *odd*.

Spreadsheet 3-1 gives a table of x- and y-values, along with a graph of $y = \dfrac{1}{x}$.

SPREADSHEET 3-1

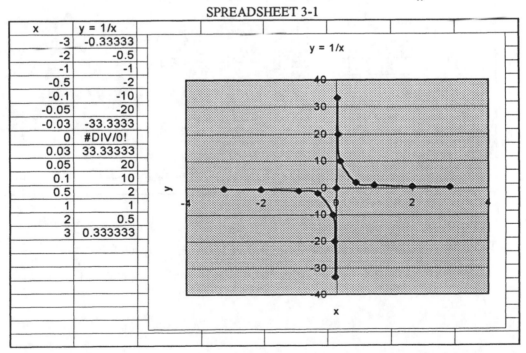

x	y = 1/x
-3	-0.33333
-2	-0.5
-1	-1
-0.5	-2
-0.1	-10
-0.05	-20
-0.03	-33.3333
0	#DIV/0!
0.03	33.33333
0.05	20
0.1	10
0.5	2
1	1
2	0.5
3	0.333333

Observe in Spreadsheet 3-1 that the y-value corresponding to $x = 0$ is given as **#DIV/0!** to indicate that division by 0 is impossible. Note that the y-values get larger and larger while getting closer to the y-axis as the x-values approach 0. Thus, the y-axis is called a ***vertical asymptote***. *Vertical asymptotes occur at values of x for which the denominator of a rational function equals 0 while the numerator does not equal 0.* The graph consists of two disconnected branches separated by the vertical asymptote, despite the fact that Excel erroneously connects these two branches with a straight line. All software packages have disadvantages, and this is one of Excel's. This underscores the need for people to learn and possess knowledge; all technologies have limitations and therefore must be controlled by human minds. Thus, if we clear the contents of cells A8 and B8 containing 0 and **#DIV/0!**, respectively, we get the graph of Spreadsheet 3-1A.

SPREADSHEET 3-1A

x	y = 1/x
-3	-0.33333
-2	-0.5
-1	-1
-0.5	-2
-0.1	-10
-0.05	-20
-0.03	-33.3333
0.03	33.33333
0.05	20
0.1	10
0.5	2
1	1
2	0.5
3	0.333333

The chart of Spreadsheet 3-1A more appropriately portrays the graph of $y = \dfrac{1}{x}$ as consisting of two disconnected branches separated by the vertical asymptote, the y-axis. Again we note that the *graph does not touch the vertical asymptote.*

48

Note also in Spreadsheet 3-1A that the y-values corresponding to negative x-values are negative, and this is why the left branch of the graph lies below the x-axis. This is true for *odd powers* of $y = \dfrac{a}{x^n}$, where a is positive.

Finally, note that the y-values approach 0 as the x-values get larger and larger in magnitude. This is why the graph approaches the x-axis for x-values of large magnitude. Accordingly, the x-axis is called a ***horizontal asymptote***. Also we note that the *graph does not touch the horizontal asymptote.*

Even Powers

The graph of $y = \dfrac{1}{x^2}$ typifies the graphs of $y = \dfrac{a}{x^n}$, where a is positive and n is ***even***.

Spreadsheet 3-2 gives a table of x- and y-values, along with a graph of $y = \dfrac{1}{x^2}$.

SPREADSHEET 3-2

x	y = 1/x^2
-2	0.25
-1.5	0.444444
-1	1
-0.5	4
-0.25	16
-0.1	100
0.1	100
0.25	16
0.5	4
1	1
1.5	0.444444
2	0.25

Observe in Spreadsheet 3-2 that we have not included $x = 0$ among the x-values. This, and the fact that after selecting **XY(Scatter)** in Excel we clicked the third graph in the first column, has resulted in a graph that appropriately has two separate branches. As with the graph of $y = \dfrac{1}{x}$, the y-axis is a *vertical asymptote* and the x-axis is a *horizontal asymptote*.

Note that both branches of the graph lie above the x-axis because the y-values are positive for all nonzero x-values as a result of the even power of x. Again we note that the *graph does not touch either the horizontal or the vertical asymptotes.*

Sums of Functions

We study the graph of $y = x + \dfrac{1}{x}$ on the interval $x > 0$ by noting that it is a sum of two

component functions, $y = x$ and $y = \dfrac{1}{x}$. Insight into its graph is gained by studying the

graphs of its component functions. Spreadsheet 3-3 gives tables of x- and y-values of the

component functions, along with a graph of the sum function, $y = x + \dfrac{1}{x}$.

SPREADSHEET 3-3

x	y = x	y = 1/x	y=x+1/x						
0.05	0.05	20	20.05						
0.25	0.25	4	4.25						
0.5	0.5	2	2.5						
0.75	0.75	1.333333	2.083333						
1	1	1	2						
2	2	0.5	2.5						
3	3	0.333333	3.333333						
4	4	0.25	4.25						
5	5	0.2	5.2						
6	6	0.166667	6.166667						

Observing the graphs in Spreadsheet 3-3, note that the straight line is the graph of $y = x$,

whereas the upper curve is the graph of the sum function, $y = x + \dfrac{1}{x}$. Note that its right

branch approaches the graph of $y = x$ for large x-values. Accordingly, the line $y = x$ is

called an *oblique asymptote*. The fact that the y-values of $y = x + \dfrac{1}{x}$ approach the line

$y = x$ can be witnessed by scanning the columns of y-values of $y = x + \dfrac{1}{x}$ and $y = x$ in Spreadsheet 3-3 and noting that for larger and larger x-values, the y-values of the sum function approach those of $y = x$, the oblique asymptote. Although the oblique asymptote, $y = x$, is included in the chart as a landmark, it is not a part of the graph of $y = x + \dfrac{1}{x}$. The graph of $y = x + \dfrac{1}{x}$ consists only of the curved line. Finally, note that the y-axis is a vertical asymptote for the sum function, $y = x + \dfrac{1}{x}$.

INSTRUCTIONS FOR SPREADSHEET 3-3

To create the graph of Spreadsheet 3-3, we highlighted all four columns of data but deleted the column labeled $y = 1/x$ in order to provide a chart containing only two graphs. To delete column labeled $y = 1/x$, during the process of creating a graph of the data, select **Source Data** and **Series**. Then in the white box labeled *Series*, click on the equation **y = 1/x** to highlight it, and click **Remove** to remove its graph from the chart.

EXERCISES

1. *Odd powers.* Create a table of x- and y-values for $y = \dfrac{1}{x^3}$.

    (a) Use the x-values -2, -1, -0.5, -0.1, -0.05, .05, 0.1, 0.5, 1, 2.
    (b) Use Chart Wizard to create the corresponding graphs.
    (c) <u>Pencil and Paper Exercise</u>. Identify the vertical and horizontal asymptotes.

    (d) <u>Pencil and Paper Exercise</u>. Does the graph's appearance resemble that of $y = \dfrac{1}{x}$ given in Spreadsheet 3-1?

2. *Even powers.* Create a table of x- and y-values for $y = \dfrac{1}{x^4}$.

    (a) Use the x-values -2, -1, -0.5, -0.1, -0.05, 0.05, 0.1, 0.5, 1, 2.
    (b) Use Chart Wizard to create the corresponding graphs.
    (c) <u>Pencil and Paper Exercise</u>. Identify the vertical and horizontal asymptotes.

    (d) <u>Pencil and Paper Exercise</u>. Does the graph's appearance resemble that of $y = \dfrac{1}{x^2}$ given in Spreadsheet 3-2?

3. *Effect of the constant a in* $y = \dfrac{a}{x^n}$. Create a table of x- and y-values for both $y = \dfrac{1}{x}$ and $y = \dfrac{5}{x}$.

    (a) Use the x-values -3, -2, -1, -0.5, -0.1, -0.05, 0.05, 0.1, 0.5, 1, 2, 3.

(b) Use Chart Wizard to create the corresponding graphs.
(c) <u>Pencil and Paper Exercise</u>. Scan the table of y-values for both equations and state the equation whose y-values are larger in magnitude and identify its graph.

4. **Effect of the constant a in** $y = \dfrac{a}{x^n}$. Create a table of x- and y-values for both

$$y = \frac{1}{x^2} \text{ and } y = \frac{7}{x^2}.$$

(a) Use the x-values -3, -2, -1, -0.5, -0.1, -0.05, 0.05, 0.1, 0.5, 1, 2, 3.
(b) Use Chart Wizard to create the corresponding graphs.
(c) <u>Pencil and Paper Exercise</u>. Scan the table of y-values for both equations and state the equation whose y-values are larger in magnitude and identify its graph.

5. **Reflection in the x-axis.** Create a table of x- and y-values for $y = \dfrac{-7}{x^2}$.

(a) Use the x-values -3, -2, -1, -0.5, -0.1, -0.05, 0.05, 0.1, 0.5, 1, 2, 3.
(b) Use Chart Wizard to create the corresponding graphs.

(c) <u>Pencil and Paper Exercise</u>. Compare this graph to that of $y = \dfrac{7}{x^2}$ in Exercise 4 and

state the effect of the negative sign before the constant multiplier, 7.

6. **Vertical shift.** Create a table of x- and y-values for both $y = \dfrac{6}{x^2}$ and $y = \dfrac{6}{x^2} + 10$.

(a) Use the x-values -3, -2, -1, -0.5, 0.5, 1, 2, 3.
     Assuming the first x-value is in cell A2, enter the formula =6/A2^2 in cell B2 and the formula =B2+10 in cell C2. State another formula that could be entered in cell C2.
(b) Use Chart Wizard to create the corresponding graphs.
(c) <u>Pencil and Paper Exercise</u>. Scan the table of y-values for both equations and state the equation whose y-values are larger in magnitude and identify its graph.
(a) <u>Pencil and Paper Exercise</u>. Identify the vertical and horizontal asymptotes.

7. <u>Pencil and Paper Exercises</u>. Draw the graph of each of the following. Identify the vertical and horizontal asymptotes for each graph.

(a) $y = \dfrac{1}{x^6}$ (b) $y = \dfrac{1}{x^5}$ (c) $y = \dfrac{4}{x}$ (d) $y = \dfrac{-4}{x}$ (e) $y = \dfrac{3}{x^2}$

(f) $y = \dfrac{-3}{x^2}$ (g) $y = \dfrac{5}{x^2} + 3$ (h) $y = \dfrac{5}{x^2} - 3$ (I) $y = \dfrac{-5}{x^2} + 3$

(j) $y = 4 + \dfrac{2}{x}$ (k) $y = 4 - \dfrac{2}{x}$ (l) $y = 8 + \dfrac{7}{x^3}$ (m) $y = 9 - \dfrac{3}{x^4}$

8. *Sum of functions.* Create a table of x- and y-values for $y = 5x + \dfrac{2}{x}$ on the interval

$x > 0$. Include tables of y-values for the component functions $y = 5x$ and $y = \dfrac{2}{x}$ as

illustrated in Spreadsheet 3-3.

(a) Use the x-values 0.1, 0.25, 0.5, 1, 2, 3, 4, 5, 6, 7, 8, 9.
   Assuming the first x-value is in cell A2, enter the formula =5*A2 in cell B2, the
   formula =2/A2 in cell C2, and the formula =B2+C2 in cell D2. State another formula
   that could be entered in cell D2.
(b) Use Chart Wizard to create the corresponding graphs.
(c) Pencil and Paper Exercise. Write the equation of the graph whose horizontal
   asymptote is the x-axis.
(d) Pencil and Paper Exercise. Write the equation of the oblique asymptote.

(e) Pencil and Paper Exercise. Identify the graph of $y = 5x + \dfrac{2}{x}$ and state its vertical

asymptote. Scan the appropriate columns of y-values to verify that the y-values of the
sum function approach those of the oblique asymptote for larger x-values.

(f) Compare the graph of $y = 5x + \dfrac{2}{x}$ with that of $y = x + \dfrac{1}{x}$ in Spreadsheet 3-3. Are

they similar in appearance?

9. *Sum of functions.* Create a table of x- and y-values for $y = 7x + \dfrac{3}{x}$ on the interval

$x > 0$. Include tables of y-values for the component functions $y = 7x$ and $y = \dfrac{3}{x}$ as

illustrated in Spreadsheet 3-3.

(a) Use the x-values 0.1, 0.25, 0.5, 1, 2, 3, 4, 5, 6, 7, 8, 9, 10.
(b) Use Chart Wizard to create the corresponding graphs.
(c) Pencil and Paper Exercise. Write the equation of the graph whose horizontal
   asymptote is the x-axis.
(d) Pencil and Paper Exercise. Write the equation of the oblique asymptote.

(e) Pencil and Paper Exercise. Identify the graph of $y = 7x + \dfrac{3}{x}$ and state its vertical

asymptote. Scan the appropriate columns of y-values to verify that the y-values of the
sum function approach those of the oblique asymptote for larger x-values.

(f) Compare the graph of $y = 7x + \dfrac{3}{x}$ with that of $y = x + \dfrac{1}{x}$ in Spreadsheet 3-3 and

also with that of $y = 5x + \dfrac{2}{x}$ in Exercise 8. Are they similar in appearance?

Summary. *(Graphs of the form* $y = ax + \dfrac{b}{x}$, *where* $x > 0$ *and* a *and* b *are positive constants)*

Exercises 8 and 9 involved graphs of equations of the form stated in the parentheses above. From these exercises, we learned that the graph of any such equation has an *oblique asymptote* given by the equation $y = ax$ and a *vertical asymptote* consisting of the y-axis.

Also, we have learned that the graph of any such equation resembles that of $y = x + \dfrac{1}{x}$ given in Spreadsheet 3-3.

10. Pencil and Paper Exercises. Draw the graph of each of the following. Identify the vertical asymptote. Write the equation of the oblique asymptote for each graph.

(a) $y = 2x + \dfrac{5}{x}$ (b) $y = 4x + \dfrac{2}{x}$ (c) $y = 4x + \dfrac{5}{x}$ (d) $y = 9x + \dfrac{4}{x}$

11. *Vertical shift.* Create a table of x- and y-values for $y = 7x + \dfrac{3}{x}$ and

$y = 5 + 7x + \dfrac{3}{x}$ on the interval $x > 0$.

    (a) Use the x-values 0.1, 0.25, 0.5, 1, 2, 3, 4, 5, 6, 7, 8, 9, 10.
        Assuming the first x-value is in cell A2, enter the formula =7*A2+3/A2 in cell B2
        and the formula =B2+5 in cell C2. State another formula that could be entered in cell
        C2.
    (b) Use Chart Wizard to create the corresponding graphs.
    (c) Pencil and Paper Exercise. Write the equation of the higher graph.
    (d) Pencil and Paper Exercise. State the relationship between the y-values of these
    graphs.
    (e) Pencil and Paper Exercise. Write the equation of the oblique asymptote for each
    graph.
    (f) Pencil and Paper Exercise. Write the equation of the vertical asymptote for each
    graph.

12. *Vertical shift.* Create a table of x- and y-values for $y = 4x + \dfrac{2}{x}$ and

$y = 7 + 4x + \dfrac{2}{x}$ on the interval $x > 0$.

    (a) Use the x-values 0.1, 0.25, 0.5, 1, 2, 3, 4, 5, 6, 7, 8, 9, 10.
    (b) Use Chart Wizard to create the corresponding graphs.
    (c) Pencil and Paper Exercise. Write the equation of the higher graph.
    (d) Pencil and Paper Exercise. State the relationship between the y-values of these
    graphs.
    (e) Pencil and Paper Exercise. Write the equation of the oblique asymptote for each
    graph.

(f) Pencil and Paper Exercise. Write the equation of the vertical asymptote for each graph.

13. Pencil and Paper Exercises. Draw the graph of each of the following. Identify the vertical asymptote. Write the equation of the oblique asymptote for each graph.

(a) $y = 8 + 5x + \dfrac{2}{x}$ (b) $y = 5 + 2x + \dfrac{3}{x}$ (c) $y = 2 + 7x + \dfrac{1}{x}$

Note: Equations of the forms presented in Exercises 8 – 13 occur in many optimization problems encountered in a study of calculus.

3-2 Average Cost Function; Inventory Cost

Average Cost Function
If $C(x)$ denotes the total cost of producing x units of some product, then the ***average cost per unit*** is given by

$$A(x) = \text{total production cost / number of units produced}$$

or, in other words,

$$A(x) = \frac{C(x)}{x} \qquad (x > 0).$$

The function $A(x)$ is called the **average cost function**. For example, if we consider the *linear cost function*

$$C(x) = 5x + 10,$$

then the *average cost function* is given by

$$A(x) = \frac{5x + 10}{x} \qquad (x > 0).$$

Spreadsheet 3-4 contains a table of x- and y-values, along with the corresponding graph of the above average cost function.

SPREADSHEET 3-4

x	A=(5x+10)/x
0	
1	15
2	10
3	8.333333
4	7.5
5	7
6	6.666667
7	6.428571
8	6.25
9	6.111111
10	6
11	5.909091
12	5.833333
13	5.769231
14	5.714286
15	5.666667

Observing the graph and table of x- and y-values in Spreadsheet 3-4, note that the x-values begin with 0 despite the fact that this x-value has no corresponding y-value. Beginning with 0 gives a graph that more accurately displays the fact that the y-axis is a *vertical asymptote* for the average cost function. After you type in 0, the formula for the y-values begins at x = 1. Follow this procedure when creating tables of x- and y-values and corresponding graphs for the spreadsheet exercises at the end of this section.

Scanning the column of y-values in Spreadsheet 3-4, note that the average cost per unit approaches the variable cost per unit, 5, as the number of units produced, x, gets larger and larger. Accordingly, the line y = 5 appears as a *horizontal asymptote* in the graph of Spreadsheet 3-4. This is confirmed algebraically by returning to the equation of the average cost function

$$A(x) = \frac{5x + 10}{x} \qquad (x > 0)$$

and dividing each term of the numerator by x to obtain the equivalent expression for $A(x)$ given below.

$$A(x) = 5 + \frac{10}{x} \qquad (x > 0)$$

The above equation reveals that the graph of $A(x)$ is obtained by beginning with the right branch of the graph of the basic form $y = \dfrac{10}{x}$ and *lifting it vertically by 5 units.* As we have already learned, this is called a *vertical shift*.

Inventory Cost

As an illustrative example, we consider a distributor selling 100,000 tires per year. For each order that the distributor places for tires, the cost to the distributor is $20. Additionally, it costs the distributor $4 to store a tire in inventory for a year. The distributor needs to determine the *number of tires that should be ordered each time an order is placed and how frequently to place such orders so that the total annual inventory cost is minimized.*

We let x denote the number of tires ordered each time the distributor places an order for tires and begin by computing the annual *ordering cost*. The annual ordering cost is determined by multiplying the cost of placing an order by the number of orders placed during a year. Because 100,000 tires are ordered in quantities of x tires per order, the number of orders placed per year is 100,000/x. The cost of placing each order is $20, so the cost of placing 100,000/x orders is 20(100,000/x) = 2,000,000/x.

Next, we determine the annual *storage cost*. This is the product of the cost of storing one tire in inventory for one year and the average number of tires in inventory. The average inventory is (0 + x)/2 = x/2, the inventory level varies from 0 to x. Because it costs $4 to store one tire in inventory for one year, the annual cost of storing x/2 tires is 4(x/2) = 2x.

Thus, the total annual inventory cost, C, is given by

$$C = \text{storage cost} + \text{ordering cost}$$

or, equivalently,

$$C(x) = 2x + \frac{2,000,000}{x} \qquad \text{where } x > 0.$$

Note that the total annual inventory cost is a function of x, the number of tires ordered with each order, and our objective is to determine x so that C is minimized.

Although this type of problem is usually solved by calculus methods, our objective here is to shed light upon the properties of the inventory cost function so that we have greater intuition when we encounter this type of problem in calculus. Accordingly, in Spreadsheet 3-5 we present a table of x- and y-values for the inventory cost function, along with its graph.

SPREADSHEET 3-5

X	C=2x+2,000,000/x
200	10400
400	5800
600	4533.333
800	4100
1000	4000
1200	4066.667
1400	4228.571
1600	4450
1800	4711.111
2000	5000
4000	8500

Observing the graph in Spreadsheet 3-5, it can be determined that the annual inventory cost is minimized when 1000 tires are ordered (i.e., at $x = 1000$) and the minimum annual inventory cost is $4000. However, more accurate methods of determining minimum (or maximum) points are studied in calculus.

As to how frequently orders should be placed, the distributor must order 100,000 tires throughout the year in lots of size 1000 tires so that $100,000/1000 = 100$ orders will be placed throughout the year.

Also, note that the inventory cost function, $C(x) = 2x + \dfrac{2,000,000}{x}$, on the interval

58

$x > 0$, is of the form $y = ax + \dfrac{b}{x}$ discussed in Exercises 8 through 10 of Section 3-1.

Using the information gained from these exercises, we determine that the straight line $y = 2x$ is an *oblique asymptote* for the cost function of Spreadsheet 3-5.

EXERCISES

1. *Total cost versus average cost per unit.* Consider the cost function $C(x) = 5x + 3$, where x denotes the number of units produced. Spreadsheet 3-6 reveals the y-values for both *total* and *average* cost, along with the graph of the average cost function.

SPREADSHEET 3-6

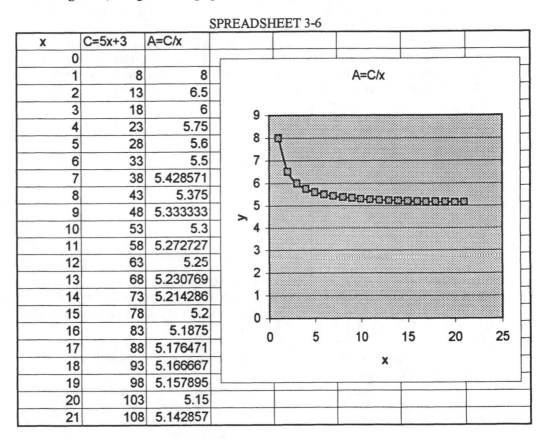

x	C=5x+3	A=C/x
0		
1	8	8
2	13	6.5
3	18	6
4	23	5.75
5	28	5.6
6	33	5.5
7	38	5.428571
8	43	5.375
9	48	5.333333
10	53	5.3
11	58	5.272727
12	63	5.25
13	68	5.230769
14	73	5.214286
15	78	5.2
16	83	5.1875
17	88	5.176471
18	93	5.166667
19	98	5.157895
20	103	5.15
21	108	5.142857

INSTRUCTIONS FOR SPREADSHEET 3-6

After selecting **XY(Scatter)** and clicking on the second graph in the first column of Chart sub-type, click **Next** and click on **Series** at the top of the box. Highlight **C=3x+5** in the Series box and click on **Remove** at the bottom of the box to remove the cost column from the graph. This will result in a better graph because otherwise the inclusion of the cost column will distort the y-axis scale. Click **Next** and finish the graph in the usual manner.

(a) <u>Pencil and Paper Exercise</u>. Verify the total costs and average costs per unit corresponding to $x = 1, 2$, and 3.

(b) <u>Pencil and Paper Exercise</u>. Write two versions of the equation of the average cost function. Which version reveals the horizontal asymptote? Write the equation of the horizontal asymptote.

(c) <u>Pencil and Paper Exercise</u>. Write two formulas that could be entered in cell C3 of Spreadsheet 3-6.

(d) <u>Pencil and Paper Exercise</u>. Scan the average cost column of y-values in Spreadsheet 3-6. The average cost per unit approaches what value as the number of units produced, x, gets larger and larger? Does this value appear to be the *horizontal asymptote* in the graph of Spreadsheet 3-6? Write the equation of part (b) that confirms this result.

2. *Average cost function.* Consider the cost function $C(x) = 3x + 4$, where x denotes the number of units produced.

(a) <u>Pencil and Paper Exercise</u>. Write the formula for the average cost function.

(b) Use the x-values 1, 2, 3, . . . 20 to create a table of x- and y-values for the average cost function. Assuming that cell A2 contains $x = 0$ and that column B contains the y-values of the average cost function, give two formulas for y-values to be entered in cell B3.

(c) Use Chart Wizard to create the graph of the average cost function.

(d) <u>Pencil and Paper Exercise</u>. Write the formula for the horizontal asymptote of the average cost function.

(e) The average cost per unit approaches what value as the number of units produced, x, gets larger and larger?

3. *Average cost function.* Consider the cost function $C(x) = x^2 - 60x + 1600$, where x denotes the number of units produced.

(a) <u>Pencil and Paper Exercise</u>. Write two versions of the formula for the average cost function.

(b) Use the x-values 10, 20, 30, 40, 50, 60, 70, 80, 90, 100, 110, 120, 130, and 140 to create a table of x- and y-values for the average cost function. Assuming that cell A2 contains $x = 0$ and that column B contains the y-values of the average cost function, give two formulas for y-values to be entered in cell B3.

(c) Use Chart Wizard to create the graph of the average cost function.

(d) <u>Pencil and Paper Exercise</u>. State what type of asymptote this average cost function has, and write its equation.

(e) <u>Pencil and Paper Exercise</u>. Scan the tables and study the graph to determine the minimum average cost and the number of units that must be produced in order to achieve the minimum average cost.

4. *Average cost function.* Consider the cost function $C(x) = x^2 - 80x + 3600$, where x denotes the number of units produced.

(a) <u>Pencil and Paper Exercise</u>. Write two versions of the formula for the average cost function.

(b) Use the x-values 10, 20, 30, 40, 50, 60, 70, 80, 90, 100, 110, 120, 130, . . . 200 to create a table of x- and y-values for the average cost function. Assuming that cell A2 contains $x = 0$ and that column B contains the y-values of the average cost function, give two formulas for y-values to be entered in cell B3.

(c) Use Chart Wizard to create the graph of the average cost function.

(d) <u>Pencil and Paper Exercise</u>. Besides the vertical asymptote, state what type of asymptote this average cost function has, and write its equation.

(e) <u>Pencil and Paper Exercise</u>. Scan the tables and study the graph to determine the minimum average cost and the number of units that must be produced in order to achieve the minimum average cost.

5. ***Inventory cost.*** If inventory is ordered in batches of x units per order, the total annual inventory cost is given by $C(x) = 8x + \dfrac{8,000,000}{x}$, where $x > 0$.

(a) Use the x-values 200, 400, 600, 800, 1000, 1200, 1400, 1600, 1800, 2000, 2200, and 2400 to create a table of x- and y-values for $C(x)$.

(b) Use Chart Wizard to create the corresponding graph.

(c) <u>Pencil and Paper Exercise</u>. Besides the vertical asymptote, state what type of asymptote this inventory cost function has, and write its equation.

(d) <u>Pencil and Paper Exercise</u>. Scan the tables and study the graph to determine the minimum inventory cost and order size, x, needed to achieve the minimum inventory cost.

6. ***Inventory cost.*** If inventory is ordered in batches of x units per order, the total annual inventory cost is given by $C(x) = 10x + \dfrac{2,500,000}{x}$, where $x > 0$.

(a) Use the x-values 100, 200, 300, 400, 500, 600, 700, 800, 900, 1000, 1100, and 1200 to create a table of x- and y-values for $C(x)$.

(b) Use Chart Wizard to create the corresponding graph.

(c) <u>Pencil and Paper Exercise</u>. Besides the vertical asymptote, state what type of asymptote this inventory cost function has, and write its equation.

(d) <u>Pencil and Paper Exercise</u>. Scan the tables and study the graph to determine the minimum inventory cost and order size, x, needed to achieve the minimum inventory cost.

CHAPTER FOUR

Matrices and Linear Systems

4-1 Matrix Operations

Often, the need arises to manipulate quantities of data that are arranged into rows and columns. Matrix operations are helpful in such situations. As examples, we use the matrices

$$A = \begin{bmatrix} 1 & 3 & 4 \\ 2 & 0 & 1 \\ -2 & 5 & 8 \end{bmatrix} \quad \text{and} \quad B = \begin{bmatrix} 3 & 2 & 0 \\ -2 & 6 & 3 \\ 2 & 4 & 1 \end{bmatrix}$$

Spreadsheet 2-1 illustrates matrices A, B, their sum $A + B$, matrix $A - 2B$, the product matrix AB, and the product matrix BA.

SPREADSHEET 4-1

	Matrix A				Matrix B	
1	3	4		3	2	0
2	0	1		-2	6	3
-2	5	8		2	4	1
	Matrix A + B				Matrix A - 2B	
4	5	4		-5	-1	4
0	6	4		6	-12	-5
0	9	9		-6	-3	6
	Matrix AB				Matrix BA	
5	36	13		7	9	14
8	8	1		4	9	22
0	58	23		8	11	20

INSTRUCTIONS

Use the following instructions to perform the matrix operations illustrated in Spreadsheet 4-1.

1. Type Labels and Matrix Entries
Move to cell B1 and type **Matrix A** to label matrix A that is to be entered below. Repeat by moving to cell F1 and typing **Matrix B** to label matrix B that is to be entered below. Note that the labels are not necessary for Excel but are helpful to us. Type the entries for matrices A and B as they appear in Spreadsheet 4-1.

2. Create the Sum Matrix A + B
2.1 Move to cell B6 and type **Matrix A + B** to label matrix A + B.
Note: Both matrices must be the same size, and the resulting sum matrix is determined by adding the corresponding entries of the two matrices. Of course, the sum matrix will be the same size as the individual matrices.

2.2 Using the mouse, move the pointer to the cell (in this example, cell A7) that will be the upper left-hand corner of the resulting sum matrix A + B. Use the mouse to highlight the cells (in this example, cells A7 through C9) that are to contain the sum matrix A + B. The highlighted cells should be black with the upper left-hand corner cell remaining white.

2.3 Type = to indicate that you are entering a formula. The = sign should automatically appear in the white cell in the upper left-hand corner.

2.4 While the highlighted cells that are to contain the sum matrix remain highlighted in black, use the mouse to move your pointer to the cell in the upper left-hand corner of matrix A and highlight the cells containing matrix A. A dotted rectangular border should appear around the selected cells. Next type + and then use your pointer to highlight the cells containing matrix B. Again, a dotted rectangular border should appear around the selected cells.

2.5 Press **Ctrl-Shift-Enter**, and the resulting sum matrix, A + B, should appear in the designated cells.

3. Create the Matrix A - 2B
3.1 Move to cell B6 and type **Matrix A - 2B** to label matrix A - 2B. Using the mouse, move the pointer to the cell (in this example, cell E7) that will be the upper left-hand corner of the resulting matrix, A - 2B. Use the mouse to highlight the cells (in this example, cells E7:G9) that are to contain the matrix A - 2B. The highlighted cells should be black with the cell in the upper left-hand corner remaining white. Note: Both matrices A and B must be the same size and matrix A - 2B is obtained by multiplying each entry of matrix B by -2 and then adding the result to matrix A. Of course, the resulting matrix A - 2B will be the same size as the individual matrices.

3.2 Type = to indicate that you are entering a formula. The = sign should automatically appear in the white cell in the upper left-hand corner.

3.3 While the highlighted cells that are to contain matrix A - 2B remain highlighted in black, use the mouse to move your pointer to the cell in the upper left-hand corner of matrix A and highlight the cells containing matrix A. Next type **-2*** and then use your pointer to highlight the cells containing matrix B. Of course, the negative sign indicates subtraction, and the symbol * indicates multiplication by a constant.

3.4 Press **Ctrl-Shift-Enter** and the resulting matrix, A - 2B, should appear in the designated cells.

4. Create the Product Matrix AB

4.1 Move to cell B11 and type **Matrix AB** to label matrix AB.
Note: The product matrix AB is defined if and only if the number of columns in the left matrix, A, equals the number of rows in the right matrix, B. Also, the product matrix AB has the same number of rows as the left matrix, A, and the same number of columns as the right matrix, B.

4.2 Using the mouse, move the pointer to the cell (in this example, cell A12) that will be the upper left-hand corner of the resulting product matrix AB. Use the mouse to highlight the cells (in this example, cells A12:C14) that are to contain the product matrix AB. The highlighted cells should be black with the cell in the upper left-hand corner remaining white.

4.3 While the highlighted cells that are to contain the product matrix remain highlighted in black, Type **=MMULT(** and use the mouse to move your pointer to the cell in the upper left-hand corner of matrix A and highlight the cells containing matrix A. Next type a comma and then use your pointer to highlight the cells containing matrix B. Type the closing parenthesis).

4.4 Press **Ctrl-Shift-Enter**, and the resulting product matrix AB should appear in the designated cells.

5. Create the Product Matrix BA

5.1 Repeat the instructions for AB, remembering that B is now the left matrix and A is the right matrix. Note: The product matrix BA is defined if and only if the number of columns in the left matrix, B, equals the number of rows in the right matrix, A. Also, the product matrix BA has the same number of rows as the left matrix, B, and the same number of columns as the right matrix, A.

5.2 Thus, when typing the formula **=MMULT(**, first highlight the cells containing matrix B, type a comma, then highlight the cells containing matrix A, and finally type the closing parenthesis).

5.3 Press **Ctrl-Shift-Enter**, and the resulting product matrix AB should appear in the designated cells.

6. Another Method for Creating a Product Matrix

6.1 After highlighting the cells that are to contain the product matrix, type =.

6.2 Use the mouse to click the f_x button from the toolbar. Move to the left column labeled **Function category** and click on **Math & Trig**. Next, move to the right column labeled **Function name** and click on **MMULT**. Click **OK**, and another dialog box appears.

6.3 Move your pointer to the white rectangle labeled **Array 1**. Click inside the rectangle and highlight the left matrix. You might have to move the dialog box in order to highlight the matrix.

6.4 Move your pointer to the white rectangle labeled **Array 2**. Click inside the rectangle and highlight the right matrix.

6.5 Press **Ctrl-Shift-Enter**. The product matrix should appear in the designated cells.

EXERCISES

1. If $A = \begin{bmatrix} -3 & 1 & 5 \\ 1 & 2 & 0 \\ 7 & 0 & -4 \end{bmatrix}$ and $B = \begin{bmatrix} 0 & 2 & -1 \\ 4 & -3 & 0 \\ 2 & 1 & -2 \end{bmatrix}$

use a spreadsheet to compute each of the following.

(a) $A + B$ (b) $A - B$ (c) $A + 2B$ (d) $A - 3B$ (e) AB (f) BA

2. *Cost analysis.* A local chain operates three fruit stores: S1, S2, and S3. Each store stocks apples, oranges, grapes, and pears. The chain does its buying on Monday, Wednesday, and Friday. Matrix M in Spreadsheet 4-2 gives the amounts spent on each item for each store on Monday. Matrices W and F give the corresponding expenditures for Wednesday and Friday, respectively.

SPREADSHEET 4-2

	Matrix M				Matrix W		
	S1	S2	S3		S1	S2	S3
Apples	200	500	300		150	80	100
Oranges	100	400	210		250	300	150
Grapes	500	280	80		70	50	90
Pears	150	350	250		120	215	160
	Matrix F						
	S1	S2	S3				
Apples	209	180	120				
Oranges	310	140	230				
Grapes	80	75	55				
Pears	95	90	170				

Use a spreadsheet to:

(a) Find the matrix that gives the combined purchases for Monday and Wednesday.

(b) Find the matrix that gives the combined purchases for Wednesday and Friday.

(c) Find the matrix that gives the combined purchases for Monday, Wednesday and Friday.

(d) Find the matrix that gives Monday's purchases plus twice Wednesday's purchases.

(e) Find the matrix that gives twice Monday's purchases minus Friday's purchases.

(f) Find the matrix that gives Monday's plus Wednesday's purchases minus Friday's purchases.

3. *Identity matrices.* Recall that an identity matrix is a square matrix I such that $AI = IA = A$, where A is a square matrix of the same dimension as I.

(a) Given the identity matrix $I = \begin{bmatrix} 1 & 0 \\ 0 & 1 \end{bmatrix}$ and matrix $A = \begin{bmatrix} 5 & -2 \\ -3 & 4 \end{bmatrix}$

use a spreadsheet to verify that $AI = A$ and $IA = A$.

(b) Given the identity matrix $I = \begin{bmatrix} 1 & 0 & 0 \\ 0 & 1 & 0 \\ 0 & 0 & 1 \end{bmatrix}$ and matrix $A = \begin{bmatrix} 2 & -1 & 5 \\ 3 & -2 & 8 \\ 7 & 2 & 1 \end{bmatrix}$

use a spreadsheet to verify that $AI = A$ and $IA = A$.

(c) Given the identity matrix $I = \begin{bmatrix} 1 & 0 & 0 & 0 \\ 0 & 1 & 0 & 0 \\ 0 & 0 & 1 & 0 \\ 0 & 0 & 0 & 1 \end{bmatrix}$ and the matrix

$$A = \begin{bmatrix} 2 & 1 & 5 & 3 \\ 7 & 2 & 4 & 1 \\ 8 & 3 & 1 & 2 \\ 3 & 2 & 5 & 4 \end{bmatrix}$$

use a spreadsheet to verify that $AI = IA = A$.

4. *Costs.* Matrix X in Spreadsheet 4-3 gives the number of units produced for each of three products during each of the indicated months. Matrix C gives the variable cost per unit for each product.

SPREADSHEET 4-3

	Matrix X					Matrix C	
	Product A	Product B	Product C			UnitVarCost	
April	200	500	300		Product A	30	
May	400	200	600		Product B	20	
June	600	800	900		Product C	50	

Use a spreadsheet to determine the matrix that gives the total combined variable cost for all three products for each of the indicated months.

5. *Production planning.* Matrix T in Spreadsheet 4-4 gives the production time (in hours) for each of two models in each of three departments. Matrix V gives the projected production volume (in units) for each model for each of the next two months, March and April.

SPREADSHEET 4-4

	Matrix T				Matrix V		
	Model A	Model B			March	April	
Dept 1	2	3		Model A	6000	8000	
Dept 2	1	2		Model B	9000	7000	
Dept 3	3	4					

Use a spreadsheet to determine the projected time requirement for each department for each of the given months.

6. *Nutrition.* Matrix A in Spreadsheet 4-5 gives the number of units of each indicated vitamin per brand of pill. Matrix N gives the number of vitamin pills of each brand that a patient takes.

SPREADSHEET 4-5

	MATRIX A					Matrix N	
	Brand 1	Brand 2	Brand 3				
Vitamin A	2	3	2		Brand 1	5	
Vitamin B	5	1	6		Brand 2	4	
Vitamin C	1	4	7		Brand 3	2	

Use a spreadsheet to determine the number of units of each vitamin taken by the patient.

7. *Production scheduling.* A company manufactures three products: A, B, and C. Each product must pass through three machines: I, II, and III. Matrix A in Spreadsheet 4-6 gives the number of hours 1 unit of product requires on the respective machines. Matrix N gives the number of units of each product to be produced the next week.

SPREADSHEET 4-6

	MATRIX A					Matrix N	
	Product A	Product B	Product C				
Machine I	3	2	3		Product A	3850	
Machine II	2	4	5		Product B	4570	
Machine II	4	6	7		Product C	1505	

Use a spreadsheet to determine the total time needed on each machine during the next week for all three products combined.

4-2 Solving Linear Systems By Matrix Inverses

A linear system such as

$$x + 3y + 3z = 4$$
$$2x + 7y + 7z = 9$$
$$2x + 7y + 6z = 10$$

can be expressed in matrix form, $AX = B$, as

$$\begin{bmatrix} 1 & 3 & 3 \\ 2 & 7 & 7 \\ 2 & 7 & 6 \end{bmatrix} \begin{bmatrix} x \\ y \\ z \end{bmatrix} = \begin{bmatrix} 4 \\ 9 \\ 10 \end{bmatrix}$$

where X denotes the solution matrix. We solve for X by multiplying both sides of the matrix equation $AX = B$ by A^{-1}, the inverse of matrix A, to obtain the solution matrix

$$X = A^{-1} B$$

provided that A^{-1} exists. Recall that the multiplicative inverse of a square matrix A is that matrix that satisfies both conditions $AA^{-1} = I$ and $A^{-1}A = I$, where I denotes the identity matrix.

We solve the above linear system in Spreadsheet 4-7 by illustrating matrices A, B, and A^{-1} and the product matrix $A^{-1} B$.

SPREADSHEET 4-7

	MATRIX A				MATRIX B		
1	3	3			4		
2	7	7			9		
2	7	6			10		
	MATRIX A^-1			SOLUTION MATRIX A^-1B			
7	-3	0			1		
-2	0	1			2		
0	1	-1			-1		

INSTRUCTIONS

Use the following instructions to perform the solution procedure illustrated in Spreadsheet 4-7.

1. Type the Labels MATRIX A, MATRIX B, MATRIX A^-1, and SOLUTION MATRIX A^-1B

2. Type in the Entries for Matrices A and B

3. Create the Matrix Inverse A^{-1}

3.1 Using the mouse, move the pointer to the cell (in this example, cell A7) that will be the upper left-hand corner of the matrix inverse A^{-1}. Use the mouse to highlight the cells (in this example, cells A7:C9) that are to contain A^{-1}. The highlighted cells should be black with the upper left-hand corner cell remaining white.

3.2 Type **=MINVERSE(** and use the mouse to highlight the cells containing matrix A. Type the closing parenthesis **)**.

3.3 Press **Ctrl-Shift-Enter**, and the resulting matrix inverse should appear in the designated cells.

4. Create the Product Matrix $A^{-1} B$
Use the **MMULT** function as discussed in Section 4-1.

5. Another Method for Creating a Matrix Inverse
5.1 After highlighting the cells that are to contain the matrix inverse, type **=**.

5.2 Use the mouse to click the f_x button from the toolbar. Move to the left column labeled **Function category** and click on **Math & Trig**. Next, move to the right column labeled **Function name** and click on **MINVERSE**. Click **OK,** and another dialog box appears.

5.3 Move your pointer to the white rectangle labeled **Array**. Click inside the rectangle and highlight the matrix whose inverse is being determined. You might have to move the dialog box in order to highlight the matrix.

5.4 Click **OK** and type **Ctrl-Shift-Enter**. The matrix inverse should appear in the designated cells.

EXERCISES

1. Use a spreadsheet to determine the inverse for each of the following matrices. Multiply each matrix by its inverse both ways to verify that the inverse satisfies both conditions $AA^{-1} = I$ and $A^{-1}A = I$, where I denotes the identity matrix.

(a) $\begin{bmatrix} 3 & -1 & 1 \\ 2 & 2 & 0 \\ 0 & 1 & 2 \end{bmatrix}$
(b) $\begin{bmatrix} 1 & 4 & 5 \\ 0 & 1 & 3 \\ 0 & 1 & 4 \end{bmatrix}$
(c) $\begin{bmatrix} 3 & 2 & 1 \\ 4 & -3 & 2 \\ 2 & 4 & -3 \end{bmatrix}$

2. *Product mix*. A company makes products A and B. Each unit of product A requires 1 unit of resource 1 and 5 units of resource 2. Each unit of product B requires 2 units of resource 1 and 4 units of resource 2. If there are 40 units of resource 1 and 140 units of resource 2 available, how many units of each product should be produced if all the resources are to be used?

(a) <u>Pencil and Paper Exercise</u>. Verify that the linear system for this problem is given by

$$x + 2y = 40$$
$$5x + 4y = 140$$

where x and y denote the numbers of units of products A and B produced, respectively.

(b) Use a spreadsheet to solve the linear system.

(c) *Recomputing a solution to accommodate a change in resources*. Say the numbers of units available of resources 1 and 2 change to 31 and 107, respectively. Use a spreadsheet to solve the resulting linear system.

(d) *Determining resources needed for a changed solution*. Say the company decides to produce 30 units of product A and 15 units of product B. Use a spreadsheet to determine how many units of each resource must be made available.

3. *Production planning*. A company makes three products: A, B, and C. In addition to a fixed cost of $20,000 for all three products combined, the unit variable costs for products A, B, and C are $5, $6, and $8, respectively. The unit profits for products A, B, and C are $1, $3, and $5, respectively. For the next month, the company needs a combined total of 11,000 units for all three products. Additionally, the company has budgeted for a total cost of $89,000 and the company requires a total profit of $31,000. How many units of each product should be made in order to satisfy the stated requirements?

(a) <u>Pencil and Paper Exercise</u>. Verify that the linear system for this problem is given by

$$x + 3y + 5z = 31{,}000$$
$$5x + 6y + 8z = 69{,}000$$
$$x + y + z = 11{,}000$$

where x, y, and z denote numbers of units of products A, B, and C produced, respectively.

(b) Use a spreadsheet to solve the linear system.

(c) Say the company decides to produce 6000 units of product A, 3000 units of product B, and 7000 units of product C. Use a spreadsheet to determine the revised total cost, total profit, and total number of units produced.

4. *Investment allocation*. An investor buys a combined total of 100 shares of stock in Companies X, Y, and Z at $20, $50, and $60 per share, respectively. The number of shares of Company Y purchased equals the combined total of the numbers of shares of Companies X and Z. If the investor invested $4700, how many shares of each company did she purchase?

(a) <u>Pencil and Paper Exercise</u>. Verify that the linear system for this problem is given by

$$x + y + z = 100$$

$$20x + 50y + 60z = 4700$$

$$-x + y - z = 0$$

where x, y, and z denote number of shares purchased of Companies A, B, and C, respectively.

(b) Use a spreadsheet to solve the linear system.

(c) The investor purchased 70, 150, and 80 shares of Companies X, Y, and Z, respectively. Use a spreadsheet to determine the total number of shares purchased and the total amount of money invested.

5. *Refinery production*. An oil company has four refineries that produce engine oil, transmission fluid, and power steering fluid. The following table gives refinery production of these three products.

Refinery Production
(in hundreds of gallons per hour)

| | Refinery | | | |
	A	B	C	D
Engine oil	2	5	4	7
Transmission fluid	1	6	3	2
Power steering fluid	8	5	4	8

The following table gives weekly orders and available hours for the next two weeks.

Weekly Orders
(in hundreds of gallons)

	Week 1	Week 2
Engine oil	420	360
Transmission fluid	340	230
Power steering fluid	550	620
Available hours (all refineries combined)	100	90

How many hours should each refinery operate each week in order to fill the weekly orders and use all available hours? Hint: There are two linear systems— one for each week.

(a) <u>Pencil and Paper Exercise</u>. Verify that the linear system for Week 1 is given by

$$2A + 5B + 4C + 7D = 420$$
$$A + 6B + 3C + 2D = 340$$
$$8A + 5B + 4C + 8D = 550$$
$$A + B + C + D = 100$$

where A, B, C, and D denote the weekly numbers of operating hours of Refineries A, B, C, and D, respectively.

(b) <u>Pencil and Paper Exercise</u>. Verify that the linear system for Week 2 is given by

$$2A + 5B + 4C + 7D = 360$$
$$A + 6B + 3C + 2D = 230$$
$$8A + 5B + 4C + 8D = 620$$
$$A + B + C + D = 90$$

where A, B, C, and D denote the weekly number of operating hours of Refineries A, B, C, and D, respectively.

(c) Use a spreadsheet to solve the linear system for Week 1.
(d) Use a spreadsheet to solve the linear system for Week 2.

6. *Nutrition*. A doctor requires that a patient take 26, 20, and 50 units of Vitamins A, B, and C, respectively, each day. The patient is to fulfill these requirements by taking a combination of three brands of pills. The following table gives the number of units of each vitamin per brand of pill.

Units of Vitamin per Brand of Pill

	Brand 1	Brand 2	Brand 3
Vitamin A	3	2	4
Vitamin B	5	1	2
Vitamin C	8	4	6

How many vitamin pills of each brand should the patient use daily in order to fulfill the doctor's requirements exactly?

(a) <u>Pencil and Paper Exercise</u>. Verify that the linear system for this problem is given by

$$3x + 2y + 4z = 26$$
$$5x + y + 2z = 20$$
$$8x + 4y + 6z = 50$$

where x, y, and z denote the numbers of vitamin pills used daily of brands A, B, and C, respectively.

(b) Use a spreadsheet to solve the linear system.

7. *Product mix*. A company makes three products, A, B, and C, each of which utilizes raw materials and labor as given in the following table. Also given is the maximum daily availability of each resource.

	Utilizations Per Unit of			Maximum daily availability
	Product A	Product B	Product C	
Raw material 1 (pounds)	2	3	0	210
Raw material 2 (pounds)	4	2	5	270
Labor (person-hours)	3	1	2	160

How many units of each product should be produced if the available resources are to be fully utilized?

(a) <u>Pencil and Paper Exercise</u>. Verify that the linear system for this problem is given by

$$2x + 3y = 210$$
$$4x + 2y + 5z = 270$$
$$3x + y + 2z = 160$$

where x, y, and z denote the numbers of units produced of products A, B, and C, respectively.

(b) Use a spreadsheet to solve the linear system.

(c) *Recomputing a solution to accommodate a change in resources*. Say the numbers of pounds of raw materials 1 and 2 available change to 160 and 210, respectively, and the number of person-hours of labor changes to 120. Use a spreadsheet to solve the resulting linear system.

(d) *Determining resources needed for a changed solution*. Say the company decides to produce 40, 75, and 35 units of products A, B, and C, respectively. Use a spreadsheet to determine how many units of each resource must be made available.

4-3 Leontief's Input–Output Model

Consider a simplified economy consisting of only three industries: A, B, and C. Leontief's model begins with a **technological** or **input–output** matrix A that gives the number of units of one industry's product needed to produce 1 unit of another industry's product as given in Spreadsheet 4-8.

SPREADSHEET 4-8

MATRIX A						
To produce 1 unit of						
P1	P2	P3				
			requires			
0.3	0.1	0.2	unit of P1			
0.2	0.1	0.1	unit of P2			
0.2	0.1	0.1	unit of P3			

The model defines a **total output** or **gross production** matrix X as

$$X = \begin{bmatrix} x_1 \\ x_2 \\ x_3 \end{bmatrix}$$

where x_1 = number of units of P1 produced, x_2 = number of units of P2 produced, and x_3 = number of units of P3 produced.

The model also defines a **consumer demand** matrix D as

$$D = \begin{bmatrix} d_1 \\ d_2 \\ d_3 \end{bmatrix}$$

where d_1 = consumer demand for P1, d_2 = consumer demand for P2, and d_3 = consumer demand for P3.

These quantities are related by the matrix equation

Total output	Internal consumption	Consumer demand

$$X \quad - \quad AX \quad = \quad D$$

The goal of input–output analysis is to determine the amount of each product, produced as given in the total output matrix X, that is needed to satisfy the final demand as given by matrix D. The solution matrix X is given by

$$X = (I - A)^{-1} D$$

assuming that $(I - A)^{-1}$ exists.

Spreadsheet 4-9 repeats the technological matrix A of Spreadsheet 4-8 while introducing a demand matrix D. The identity matrix I is given, along with matrices I - A and $(I - A)^{-1}$ and the product matrix $(I - A)^{-1}D$.

SPREADSHEET 4-9

MATRIX A				IDENTITY MATRIX I			
0.3	0.1	0.2		1	0	0	
0.2	0.1	0.1		0	1	0	
0.2	0.1	0.1		0	0	1	
MATRIX I - A				MATRIX (I - A)^-1			
0.7	-0.1	-0.2		1.6	0.22	0.38	
-0.2	0.9	-0.1		0.4	1.18	0.22	
-0.2	-0.1	0.9		0.4	0.18	1.22	
MATRIX D				SOLUTION MATRIX X=(I - A)^-1D			
	50				120.8		
	30				75.2		
	90				135.2		

The solution matrix X in Spreadsheet 4-9, shows that 120.8, 75.2, and 135.2 units of products P1, P2, and P3, respectively, must be produced in order to satisfy the demands given in matrix D.

INSTRUCTIONS

Use the following instructions to perform the matrix operations illustrated in Spreadsheet 4-9.

1. Type Matrix Labels
2. Type the Entries for Matrices A, I, and D
3. Create Matrix I - A
4. Create Matrix (I - A)$^{-1}$
5. Create Matrix (I - A)$^{-1}$D

EXERCISES

1. The technological or input–output matrix A and the consumer demand matrix D for some economy are given as follows.

$$A = \begin{bmatrix} 0.1 & 0.2 \\ 0.3 & 0.1 \end{bmatrix} \qquad D = \begin{bmatrix} 30 \\ 20 \end{bmatrix}$$

(a) Use a spreadsheet to determine the total output matrix X that gives the outputs needed to satisfy the demands given in matrix D.

(b) Use a spreadsheet to determine the internal consumption of each product.

2. The technological or input–output matrix A and the consumer demand matrix D for some economy are given as follows.

$$A = \begin{bmatrix} 0.1 & 0.4 \\ 0.2 & 0.3 \end{bmatrix} \qquad D = \begin{bmatrix} 800 \\ 1000 \end{bmatrix}$$

(a) Use a spreadsheet to determine the total output matrix X that gives the outputs needed to satisfy the demands given in matrix D.

(b) Use a spreadsheet to determine the internal consumption of each product.

3. The technological or input–output matrix A and the consumer demand matrix D for some economy are given as follows.

$$A = \begin{bmatrix} 0.1 & 0.2 & 0.2 \\ 0.2 & 0.1 & 0.2 \\ 0.2 & 0.2 & 0.1 \end{bmatrix} \qquad D = \begin{bmatrix} 100 \\ 200 \\ 180 \end{bmatrix}$$

(a) Use a spreadsheet to determine the total output matrix X that gives the outputs needed to satisfy the demands given in matrix D.

(b) Use a spreadsheet to determine the internal consumption of each product.

4. The technological or input–output matrix A and the consumer demand matrix D for some economy are given as follows.

$$A = \begin{bmatrix} 0.3 & 0.1 & 0.4 \\ 0.25 & 0.4 & 0.2 \\ 0.2 & 0.4 & 0.15 \end{bmatrix} \qquad D = \begin{bmatrix} 20 \\ 15 \\ 40 \end{bmatrix}$$

(a) Use a spreadsheet to determine the total output matrix X that gives the outputs needed to satisfy the demands given in matrix D.

(b) Use a spreadsheet to determine the internal consumption of each product.

5. The technological or input–output matrix A and the consumer demand matrix D for some economy are given as follows.

$$A = \begin{bmatrix} 0.3 & 0.2 & 0.4 & 0.1 \\ 0.2 & 0.1 & 0.2 & 0.3 \\ 0.3 & 0.1 & 0.2 & 0.3 \\ 0.4 & 0.2 & 0.1 & 0.2 \end{bmatrix} \qquad D = \begin{bmatrix} 40 \\ 60 \\ 70 \\ 20 \end{bmatrix}$$

(a) Use a spreadsheet to determine the total output matrix X that gives the outputs needed to satisfy the demands given in matrix D.

(b) Use a spreadsheet to determine the internal consumption of each product.

CHAPTER FIVE

Linear Programming

Linear programming problems involve the optimizing (maximizing or minimizing) of a linear objective function subject to linear constraints. The following is an introductory example.

A company makes two products—product 1 and product 2. Each unit of product 1 requires 1 unit of raw material A and 1 unit of raw material B. Each unit of product 2 requires 1 unit of raw material A and 2 units of raw material B. There are only 500 units of raw material A and 800 units of raw material B available. The company makes a profit of $1 and $3 on each unit of products 1 and 2, respectively. The company must determine an *optimal product mix*— in other words, how many units of each product to produce in order to *maximize total profit* and *satisfy the raw material constraints.* This problem is outlined in the table below.

	x Product 1	y Product 2	
Raw material A	1	1	≤ 500
Raw material B	1	2	≤ 800
Unit profits	1	3	

Note that we let

x = number of units of product 1 to be produced

y = number of units of product 2 to be produced.

P = total profit

Our problem is formulated as follows.

Objective function
$$\text{Maximize } P = x + 3y$$

subject to the following *constraints.*
$$x + y \leq 500$$
$$x + 2y \leq 800$$
$$x \geq 0$$
$$y \geq 0$$

The spreadsheet below, which is followed by Excel instructions, illustrates the entries for the above problem.

MAXIMIZATION PROBLEM					
Objective					
Profit	0				
Decision variables					
Product 1 (x)	0				
Product 2 (y)	0				
Constraints					
Raw material A	0	500			
Raw material B	0	800			
Nonnegativity 1	0	0			
Nonnegativity 2	0	0			

INSTRUCTIONS

1. Type Labels
Type the labels in column A as shown. When needed, _widen column A_ by moving the mouse pointer to the column heading so that it points to the _right-hand border_ of column A. When the mouse pointer becomes a two-way arrow, _double-click the right-hand border of the column_, and it should widen to accommodate the cell entries of column A.

2. Enter Formulas and Initial Values
2.1 Enter the formula =1*B9+3*B10 for the _objective (profit) function_ in cell B5. A zero will appear in the cell because cells B9 and B10 have no values at this point. Type **0** for the initial values of x and y in cells B9 and B10. As indicated, these cells contain the values of the decision variables x and y, respectively.

2.2 Enter the formula =1*B9+1*B10 for the _left-hand side of Raw material A constraint_ in cell B14 and type the _right-hand-side constant_, **500**, in cell C14. A zero will appear in cell B14.

2.3 Enter the formula =1*B9+2*B10 for the _left-hand side of Raw material B constraint_ in cell B15 and type the _right-hand-side constant_, **800**, in cell C15. A zero will appear in cell B15.

2.4 Enter the formula **=B9** for the *left-hand side of the nonegativity constraint* $x \geq 0$ in cell B16 and type the *right-hand-side constant*, **0**, in cell C16. A zero will appear in cell B16.

2.5 Enter the formula **=B10** for the *left-hand side of the nonegativity constraint* $y \geq 0$ in cell B17 and type the *right-hand-side constant*, **0**, in cell C17. A zero will appear in cell B17.

3. Use Solver

(Enter the objective function.)

3.1 Use the mouse to *move the dark-bordered rectangle to the cell containing the formula for the objective function*—in this case cell B5, which currently contains the value 0.

3.2 Select *Tools*, then choose *Solver*, and a dialog box appears with the *objective function cell*, B5, identified in the *Set Target Cell* box as **B5**. This cell contains the formula that Solver tries to maximize or minimize. Click on *Max* to maximize and click on the white box under *By Changing Cells*. Then either select the cells (using the mouse) containing the values of x and y, cells B9 and B10, or type in **B9:B10**.

(Enter the constraints.)

3.3 Click *Add* to the right of the white box appearing under *Subject to the Constraints* and a dialog box appears. Either select the cell (using the mouse), in this case B14, containing the formula for the first constraint, and click inside the white box under *Cell Reference*, or type **B14** inside the white box. Choose the appropriate inequality symbol, in this case <=, and again either select the cell (using the mouse) containing the right-hand-side constant and click inside the white box located to the right of the inequality symbol, or type **C14** inside the white box. Click *Add* to include this constraint.

3.4 Repeat this procedure for any remaining constraints, including the nonnegativity restrictions.

3.5 After you have added all the constraints, click *Cancel* and the original Solver dialog box appears with a listing of the cells corresponding to the constraints.

(Obtain the optimal solution.)

3.6 Click *Options* and use the clicker to place a check in the small white box next to *Assume Linear Model*. If the check is already there, this step is not needed. Click *OK* to return to the Solver dialog box.

3.7 Click *Solve*, and Solver determines optimal values (if they exist) for the decision variables. Note that Solver changes the values in your spreadsheet by replacing them with the optimal solution.

3.8 Use the Solver Results dialog box to prepare a report that details the results of the optimal solution on the problem. Double-click *Answer* in the *Reports* box and click *OK*.

Finally, click on *Answer report* at the bottom of your screen to obtain the report. The report for our illustrative problem is given below.

ANSWER REPORT

Target Cell (Max)

Cell	Name	Original Value	Final Value
$B$5	Profit (P)	0	1200

Adjustable Cells

Cell	Name	Original Value	Final Value
$B$9	Product 1 (x)	0	0
$B$10	Product 2 (y)	0	400

Constraints

Cell	Name	Cell Value	Formula	Status	Slack
$B$14	Raw material A	400	$B$14<=$C$14	Not Binding	100
$B$15	Raw material B	800	$B$15<=$C$15	Binding	0
$B$16	Nonnegativity 1	0	$B$16>=$C$16	Binding	0
$B$17	Nonnegativity 2	400	$B$17>=$C$17	Not Binding	400

Studying the **Target Cell** and **Adjustable Cells** sections of the Answer Report, note that the maximum profit of $1200 results from producing 0 units of product 1 and 400 units of product 2. Substituting $x = 0$ and $y = 400$ into the objective function verifies the optimal profit function value of 1200.

Studying the **Constraints** section, note that the Raw material A constraint has 100 units of slack, which indicates that only 400 of the available 500 units of Raw material A are used. The status of this constraint is appropriately termed *Not Binding*. Substituting $x = 0$ and $y = 400$ into the respective inequality verifies this result. Recall that a *binding constraint* is one where the slack (or surplus) is 0.

Note that the Raw material B constraint has 0 units of slack, which indicates that all of the available 500 units of Raw material B are used. The status of this constraint is appropriately termed *Binding*. Substituting $x = 0$ and $y = 400$ into the respective inequality verifies this result.

81

EXERCISES

1. *Product mix.* A firm produces products A, B, and C, each of which passes through assembly and inspection departments. The number of person-hours required by a unit of each product in each department is given as follows.

Person-hours per unit of product

	Product A	Product B	Product C
Assembly	2	4	2
Inspection	3	2	1

During a particular week, the assembly and inspection departments have available at most 1500 and 1200 person-hours, respectively. If the unit profits for products A, B, and C are $50, $40, and $60, respectively, determine the number of units of each product that should be produced in order to maximize the total profit and satisfy the constraints of the problem.

(a) <u>Pencil and Paper Exercise</u>. Verify that this problem is formulated as follows.

$$\text{Maximize } P = 50x_1 + 40x_2 + 60x_3$$
$$\text{subject to } \quad 2x_1 + 4x_2 + 2x_3 \leq 1500$$
$$3x_1 + 2x_2 + x_3 \leq 1200$$
$$x_1, x_2, x_3 \geq 0$$

where P denotes total profit and x_1, x_2, and x_3 denote the numbers of units of products A, B, and C, respectively, to be produced.
(b) Use a spreadsheet to solve this problem.
(c) <u>Paper and Pencil Exercise</u>. Using the spreadsheet results of part (b), state the optimal solution.
(d) <u>Pencil and Paper Exercise</u>. Verify the maximum profit by substituting the solution values into the objective function.
(e) <u>Pencil and Paper Exercise</u>. Verify the slack in each constraint by substituting the solution values into the respective inequalities.

2. *Agriculture: Crop allocation.* A farmer has 1000 acres to be allocated among three crops. The number of person-hours per acre required for each crop and its profit per acre are given as follows.

Person-hours per acre

Crop A	Crop B	Crop C
3	5	4

Profit per acre

Crop A	Crop B	Crop C
$600	$300	$500

If the farmer has at most 6000 person-hours available, determine the number of acres of each crop that should be planted in order to maximize the total profit.

(a) Pencil and Paper Exercise. Verify that this problem is formulated as follows.

$$\text{Maximize} \quad P = 600x_1 + 300x_2 + 500x_3$$
$$\text{subject to} \quad 3x_1 + 5x_2 + 4x_3 \leq 6000$$
$$x_1 + x_2 + x_3 \leq 1000$$
$$x_1, x_2, x_3 \geq 0$$

where P denotes total profit and x_1, x_2, and x_3 denote the numbers of acres of crops A, B, and C, respectively, to be planted.

(b) Use a spreadsheet to solve this problem.

(c) Paper and Pencil Exercise. Using the spreadsheet results of part (b), state the optimal solution.

(d) Pencil and Paper Exercise. Verify the maximum profit by substituting the solution values into the objective function.

(e) Pencil and Paper Exercise. Verify the slack in each constraint by substituting the solution values into the respective inequalities.

3. *Agriculture: Crop allocation.* Repeat Exercise 2 under the additional constraints that there should be no more than 500 and 300 acres of crops A and C, respectively.

(a) Pencil and Paper Exercise. Verify that this problem is formulated as follows.

$$\text{Maximize} \quad P = 600x_1 + 300x_2 + 500x_3$$
$$\text{subject to} \quad 3x_1 + 5x_2 + 4x_3 \leq 6000$$
$$x_1 + x_2 + x_3 \leq 1000$$
$$x_1 \leq 500$$
$$x_3 \leq 300$$
$$x_1, x_2, x_3 \geq 0$$

where P denotes total profit and x_1, x_2, and x_3 denote the numbers of acres of crops A, B, and C, respectively, to be planted.

(b) Use a spreadsheet to solve this problem.

(c) Paper and Pencil Exercise. Using the spreadsheet results of part (b), state the optimal solution.

(d) Pencil and Paper Exercise. Verify the maximum profit by substituting the solution values into the objective function.

(e) Pencil and Paper Exercise. Verify the slack in each constraint by substituting the solution values into the respective inequalities.

4. *Nutrition: Diet.* A dietician must determine how many units of each of three food types to include in a diet so that the total number of calories is minimized and nutritional requirements are realized. The following table gives the amounts of relevant nutrients per unit of each food type.

Milligrams per unit of food type

	Food A	Food B	Food C
Protein	5	4	2
Calcium	10	5	8
Iron	2	1	3

Calories per unit of food type

Food A	Food B	Food C
800	500	600

Minimal nutritional requirements (in milligrams)

Protein	25
Calcium	60
Iron	10

(a) <u>Pencil and Paper Exercise</u>. Verify that this problem is formulated as follows.

$$\text{Minimize } C = 800x_1 + 500x_2 + 600x_3$$
$$\text{subject to } 5x_1 + 4x_2 + 2x_3 \geq 25$$
$$10x_1 + 5x_2 + 8x_3 \geq 60$$
$$2x_1 + x_2 + 3x_3 \geq 10$$
$$x_1, x_2, x_3 \geq 0$$

where C denotes total number of calories and x_1, x_2, and x_3 denote the numbers of units of foods A, B, and C, respectively, to be included in the diet.

(b) Use a spreadsheet to solve this problem.

(c) <u>Paper and Pencil Exercise</u>. Using the spreadsheet results of part (b), state the optimal solution.

(d) <u>Pencil and Paper Exercise</u>. Verify the maximum profit by substituting the solution values into the objective function.

(e) <u>Pencil and Paper Exercise</u>. Verify the surplus in each constraint by substituting the solution values into the respective inequalities.

5. *Nutrition: Diet.* Repeat Exercise 4 with the objective of minimizing the cost of the diet. Delete the objective of minimizing the total number of calories. The unit cost of each food type is given in the following table.

Cost per unit of food type

Food A	Food B	Food C
$15	$20	$30

(a) <u>Pencil and Paper Exercise</u>. Verify that this problem is formulated as follows.

$$\text{Minimize } C = 15x_1 + 20x_2 + 30x_3$$
$$\text{subject to } 5x_1 + 4x_2 + 2x_3 \geq 25$$
$$10x_1 + 5x_2 + 8x_3 \geq 60$$
$$2x_1 + x_2 + 3x_3 \geq 10$$
$$x_1, x_2, x_3 \geq 0$$

where C denotes the total cost of the diet and x_1, x_2, and x_3 denote the numbers of units of foods A, B, and C, respectively, to be included in the diet.
(b) Use a spreadsheet to solve this problem.
(c) <u>Paper and Pencil Exercise</u>. Using the spreadsheet results of part (b), state the optimal solution.
(d) <u>Pencil and Paper Exercise</u>. Verify the maximum profit by substituting the solution values into the objective function.
(e) <u>Pencil and Paper Exercise</u>. Verify the surplus in each constraint by substituting the solution values into the respective inequalities.

6. *Investments: Portfolio selection.* An investor has $600,000 to allocate among these investment alternatives.

Investment alternatives	Rate of return	Risk level
Fixed-income funds	7.5%	0
Equity income funds	8.6	2
Aggressive growth funds	13.5	9
Balanced funds	7.8	1

The investor has the following investment criteria:
1. The full amount of $600,000 is to be invested.
2. The average risk is to be no more than 3.5.
3. At least $100,000 is to be invested in fixed-income funds.
4. At least $100,000 is to be invested in equity income funds.
5. At least $60,000 is to be invested in aggressive growth funds.
6. At least $50,000 is to be invested in balanced funds.

Determine the amount that should be invested in each investment alternative in order to maximize the annual return.
(a) <u>Pencil and Paper Exercise</u>. Verify that this problem is formulated as follows.

$$\text{Maximize } R = 0.075x_1 + 0.086x_2 + 0.135x_3 + 0.078x_4$$
$$\text{subject to } x_1 + x_2 + x_3 + x_4 = 600,000$$
$$2x_2 + 9x_3 + x_4 \leq 2,100,000$$
$$x_1 \geq 100,000$$
$$x_2 \geq 100,000$$

$$x_3 \geq 60,000$$
$$x_4 \geq 50,000$$
$$x_1, x_2, x_3, x_4 \geq 0$$

where R denotes the annual return and x_1, x_2, x_3 and x_4 denote the amounts invested in fixed-income, equity income, aggressive growth, and balanced funds, respectively.

Hint: The second constraint, which states that the average risk must be at most 3.5, is formulated as follows. Using the risk levels given in the table, x_1, x_2, x_3, and x_4 dollars will be subjected to risk levels of 0, 2, 9, and 1, respectively. Thus, the average risk level, which must be at most 3.5, is given by

$$\frac{0x_1 + 2x_2 + 9x_3 + x_4}{600,000} \leq 3.5.$$

Verify that multiplying both sides of the above inequality by 600,000 gives the second constraint

$$2x_2 + 9x_3 + x_4 \leq 2,100,000.$$

(b) Use a spreadsheet to solve this problem.
(c) <u>Paper and Pencil Exercise</u>. Using the spreadsheet results of part (b), state the optimal solution.
(d) <u>Pencil and Paper Exercise</u>. Verify the maximum profit by substituting the solution values into the objective function.
(e) <u>Pencil and Paper Exercise</u>. Verify the slack (or surplus) in each constraint by substituting the solution values into the respective inequalities.

7. *Transportation problem.* A company makes water heaters at two plants. The water heaters must be transported from the two plants to two distribution centers, as illustrated in the following table. Note that plants 1 and 2 have production capacities of 500 and 600 water heaters, respectively. Distribution centers 1 and 2 have demand requirements of 400 and 650 water heaters, respectively. The numbers in the upper left-hand corner of the cells in the table represent the cost of transporting a water heater from the corresponding plant to the respective distribution center. Also, the numbers of water heaters to be shipped from respective plants to respective distribution centers are denoted by x_1, x_2, x_3, and x_4. The company must determine how many water heaters to ship from each plant to each distribution center in order to minimize the total transportation cost.

	Distribution center 1	Distribution center 2	Plant capacities
Plant 1	$6 x_1	$8 x_2	≤ 500
Plant 2	$10 x_3	$9 x_4	≤ 600
Demands	≥ 400	≥ 650	

(a) <u>Pencil and Paper Exercise</u>. Verify that this problem is formulated as follows.

$$\text{Minimize}\quad C = 6x_1 + 8x_2 + 10x_3 + 9x_4$$
$$\text{subject to}\quad x_1 + x_2 \leq 500$$
$$x_3 + x_4 \leq 600$$
$$x_1 + x_3 \geq 400$$
$$x_2 + x_4 \geq 650$$
$$x_1, x_2, x_3, x_4 \geq 0$$

where C denotes the total transportation cost and x_1, x_2, x_3, and x_4 are as defined above.

(b) Use a spreadsheet to solve this problem.

(c) <u>Paper and Pencil Exercise</u>. Using the spreadsheet results of part (b), state the optimal solution.

(d) <u>Pencil and Paper Exercise</u>. Verify the maximum profit by substituting the solution values into the objective function.

(e) <u>Pencil and Paper Exercise</u>. Verify the slack (or surplus) in each constraint by substituting the solution values into the respective inequalities.

8. *Break-even analysis.* The unit selling price, the variable cost per unit, and the fixed cost are given in the following table for each of three products produced by a company. The company wishes to determine how many units of each product it must produce and sell in order to break even and minimize the total combined variable cost subject to the following restrictions:

- Marketing research has indicated that no more than 4000 units of product 1 should be produced.
- The company has an order for 600 units of product 2. Thus, at least 600 units of product 2 must be produced and sold.
- The company has an order for 700 units of product 3. Thus, at least 700 units of product 3 must be produced and sold.

	Selling price per unit	Variable cost per unit	Fixed cost
Product 1	$50	$30	$50,000
Product 2	80	50	30,000
Product 3	40	25	30,000

(a) <u>Pencil and Paper Exercise</u>. Verify that this problem is formulated as follows.

Minimize $C = 30x_1 + 50x_2 + 25x_3$

subject to
$$20x_1 + 30x_2 + 15x_3 = 110{,}000$$
$$x_1 \leq 4000$$
$$x_2 \geq 600$$
$$x_3 \geq 700$$
$$x_1, x_2, x_3 \geq 0$$

where C denotes the total combined variable cost and x_1, x_2, and x_3 denote the numbers of units produced and sold of products 1, 2, and 3, respectively.

Hint: The first constraint expresses the requirement that the company break even, which means that

Total sales revenue = total cost.

Verify that *total sales revenue* is given by the expression $50x_1 + 80x_2 + 40x_3$ and that the *total combined cost* is given by the expression

$$30x_1 + 50x_2 + 25x_3 + (50{,}000 + 30{,}000 + 30{,}000).$$

Equate these two expressions and simplify the result to obtain the first constraint.

(b) Use a spreadsheet to solve this problem.
(c) <u>Paper and Pencil Exercise</u>. Using the spreadsheet results of part (b), state the optimal solution.
(d) <u>Pencil and Paper Exercise</u>. Verify the maximum profit by substituting the solution values into the objective function.
(e) <u>Pencil and Paper Exercise</u>. Verify the slack (or surplus) in each constraint by substituting the solution values into the respective inequalities.

CHAPTER SIX

Mathematics of Finance

6-1 Simple Versus Compound Interest

If $20 earns simple interest at an annual rate of 10%, then the interest for 1 year is determined by multiplying $20 by 10% or, equivalently, $(20)(0.10) = \$2$. This means that the $20 investment increases in value by $2 per year. If y denotes this investment's value after x years, then y and x are related by the *linear equation*

$$y = 20 + 2x.$$

If $20 earns interest at 10% compounded annually, this means that at the end of the first year, the $20 is increased by 10% of itself to become $20(1.10)$; at the end of the second year, the $20(1.10)$ is increased by 10% of itself to become $20(1.10)(1.10)$, or $20(1.10)^2$; at the end of the third year, the $20(1.10)^2$ is increased by 10% of itself to become $20(1.10)^2(1.10)$, or $20(1.10)^3$. Thus, if y denotes this investment's value after x years, then y and x are related by the *exponential equation*

$$y = 20(1.10)^x.$$

Spreadsheet 6-1 gives a table of x- and y-values for both equations.

SPREADSHEET 6-1

x	y=20+2x	y=20(1.10)^x
0	20	20
1	22	22
2	24	24.2
3	26	26.62
4	28	29.282
5	30	32.2102
6	32	35.43122
7	34	38.974342
8	36	42.8717762
9	38	47.15895382
10	40	51.8748492

Observing the table and graphs in Spreadsheet 6-1, note that simple interest results in linear growth, whereas compound interest results in exponential growth. Specifically, move down the column of y-values of the simple interest equation, $y = 20 + 2x$, and note that the value of the investment increases by the same amount each year, whereas the y-values growing at compound interest increase by larger and larger amounts each year. Observe how the graph of the exponential (compound interest) equation pulls away to the up side from the graph of the linear (simple interest) equation. Clearly the y-values for compound interest are increasing faster than their corresponding counterparts for simple interest, and that is why the compound interest curve pulls away from the simple interest line.

INSTRUCTIONS

Use the following instructions to create tables and graphs similar to those in Spreadsheet 6-1.

1. Create a Table of x- and y-values
1.1 After labeling Columns A, B, and C as **x, y = 20 + 2x**, and **y = 20(1.10)^x**, respectively, move to cell A2 and type the x-values **0** through **10** into Column A as illustrated in Spreadsheet 6-1.

1.2 *Enter the formulas* to compute the y-values for the equation $y = 20 + 2x$ in

Column B and $y = 20(1.10)^x$ in Column C. Move the dark-bordered rectangle to cell B2, type the formula **=20+2*A2**, and press **Enter**. Recall that the symbol ***** means *multiplication*. Next, move the dark-bordered rectangle to cell C2, type the formula **=20*(1.10)^A2**, and press **Enter**. Note that the symbol **^** means *exponentiation*.

1.3 *Copy the formulas*. Begin with the formula **=20+2*A2** in cell B2 and copy it down through cell B12. Repeat the procedure for the formula **=20*(1.10)^A2** in cell C2.

2. Create a Graph of the Data
2.1 *To create a graph of the data*, move the mouse pointer to the middle of cell A1, click, hold, and drag the mouse pointer until it highlights the cells containing the data and labels. Cell A1 will remain unhighlighted.

2.2 Select Chart Wizard from the toolbar, and a dialog box appears.

Step 1: Select **XY(Scatter)** in the Chart type section. Move to the Chart sub-type section and click on the second graph in the first column. Click **Next**.

Step 2: Click **Next**.

Step 3: Enter **x** in the Value(X) axis section and **y** in the Value(Y) axis section to label the x- and y-axes. Click **Next**.

Step 4: Click **Finish**, and the graph should appear in your spreadsheet.

3. Change the Size of the Graph
Move the mouse pointer to the middle handle at the bottom of the chart until the pointer becomes a vertical double-sided arrow. Click, hold, and drag the line downward to enlarge the chart vertically. To enlarge the chart horizontally, move the mouse pointer to the middle handle at the side of the chart until the pointer becomes a horizontal double-sided arrow. Click, hold, and drag the line horizontally to enlarge the chart. Dragging the line horizontally in the reverse direction will decrease the size of the chart.

EXERCISES

1. *Simple versus compound interest.* If $2000 earns simple interest at an annual rate of 5% and y denotes the value of this investment after x years, verify that the equation that relates y and x is $y = 2000 + 100x$. If $2000 earns interest at 5% compounded annually, verify that the equation that relates y and x is $y = 2000(1.05)^x$.

(a) Use a spreadsheet to compare the simple and compound interest values of this investment as illustrated in Spreadsheet 6-1. Use the x-values 0, 1, 2, 3, 4, 5, 6, 7, 8, 9, and 10 to create tables of x- and y-values and the accompanying graphs.
(b) State the values of this investment for both simple and compound interest at the end of 5 years.
(c) State the values of this investment for both simple and compound interest at the end of 10 years.
(d) Studying the accompanying graph, indicate which graph represents growth at simple interest and which at compound interest.

2. *Simple versus compound interest.* If $5000 earns simple interest at an annual rate of 8% and y denotes the value of this investment after x years, verify that the equation that relates y and x is $y = 5000 + 400x$. If $5000 earns interest at 8% compounded annually, verify that the equation that relates y and x is $y = 5000(1.08)^x$.

(a) Use a spreadsheet to compare the simple and compound interest values of this investment as illustrated in Spreadsheet 6-1. Use the x-values 0, 1, 2, 3, 4, 5, 6, 7, 8, 9, and 10 to create tables of x- and y-values and the accompanying graphs.
(b) State the values of this investment for both simple and compound interest at the end of 5 years.
(c) State the values of this investment for both simple and compound interest at the end of 10 years.
(d) Studying the accompanying graph, indicate which graph represents growth at simple interest and which at compound interest.

3. *Effect of annual rate on investment results.* The compound amount formula $S = P(1+i)^n$ gives the compound amount, S, of a principal of P dollars after n compoundings at an interest rate i per conversion period; i is calculated by the formula

$i = \dfrac{r}{m}$, where r is the *annual interest rate* and m is the *number of compoundings in* 1 *year*. Spreadsheet 6-2 illustrates a format for comparing the compound amount of a $10,000 investment after 5 years at the indicated annual rates compounded monthly.

SPREADSHEET 6-2

Principal	AnnRate	n	CompAmt			
10000	0.04	60	12209.97			
10000	0.05	60	12833.59			
10000	0.06	60	13488.5			
10000	0.07	60	14176.25			
10000	0.08	60	14898.46			
10000	0.09	60	15656.81			
10000	0.1	60	16453.09			

The compound amounts of column D were determined by *entering the formula* =A2*(1+B2/12)^C2 in cell D2 and *copying that formula* throughout the indicated cells of column D. Note that the formula is the *compound amount formula*, where **A2** is the principal, **B2/12** is the annual interest rate divided by 12 compoundings per year (i.e., compounded monthly), and **C2** is n, the total number of compoundings.
   (a) Create Spreadsheet 6-2 on your computer.
   (b) <u>Pencil and Paper Exercise</u>. Compare the compound amounts for the annual rates of 5% and 6%, stating how much more an investor would have at the higher rate.
   (c) <u>Pencil and Paper Exercise</u>. Compare the compound amounts for the annual rates of 4% and 10%, stating how much more an investor would have at the higher rate.

4. *Effect of annual rate on investment results.* Create a spreadsheet similar to Spreadsheet 6-2 for a principal of $50,000 invested for 10 years at annual interest rates of 5%, 6%, 7%, 8%, 9%, 10%, 11%, and 12%, all compounded quarterly.
   (a) <u>Pencil and Paper Exercise</u>. Compare the compound amounts for the annual rates of 5% and 6%, stating how much more an investor would have at the higher rate.
   (b) <u>Pencil and Paper Exercise</u>. Compare the compound amounts for the annual rates of 8% and 12%, stating how much more an investor would have at the higher rate.
   (c) <u>Pencil and Paper Exercise</u>. Compare the compound amounts for the annual rates of 5% and 12%, stating how much more an investor would have at the higher rate.

5. *Graphical interpretation of annual rate.* If a principal of $5 earns interest at 30% compounded annually, its value, y, after x years is given by $y = 5(1.30)^x$, whereas at 60% compounded annually, its value, y, after x years is given by $y = 5(1.60)^x$.
   (a) Use a spreadsheet to create a table of x- and y-values, along with corresponding graphs of both equations, for the x-values 0, 1, 2, 3, 4, 5, and 6.
   (b) <u>Pencil and Paper Exercise</u>. State the equation of the steeper curve. Does this curve represent the investment that is growing more rapidly? It should be obvious that the annual rate is the engine that drives exponential growth.

92

6. *Graphical interpretation of annual rate.* If a principal of $10 earns interest at 15% compounded annually, its value, y, after x years is given by $y = 10(1.15)^x$, whereas at 40% compounded annually, its value, y, after x years is given by $y = 10(1.40)^x$

(a) Use a spreadsheet to create a table of x- and y-values, along with corresponding graphs of both equations, for the x-values 0, 1, 2, 3, 4, 5, 6 and 7.

(b) <u>Pencil and Paper Exercise</u>. State the equation of the curve representing the investment that is growing more rapidly.

7. *Effective rate.* If an investment earns interest at an annual rate r compounded m times per year, the simple interest rate that produces the same returns after 1 year is called the *effective rate* and is determined by the formula

$$EffRate = \left(1 + \frac{r}{m}\right)^m - 1.$$

The effective rate provides the common basis of a 1-year time period for comparing different interest rates.

Spreadsheet 6-3 provides a format for computing the effective rates corresponding to the annual rate of 5% compounded m times per year for the indicated values of m.

SPREADSHEET 6-3

AnnRate	m	EffRate				
0.05	1	0.05				
0.05	2	0.050625				
0.05	4	0.050945				
0.05	6	0.051053				
0.05	12	0.051162				
0.05	365	0.051267				

The effective rates of Column C were determined by *entering the formula* **=(1+A2/B2)^B2-1** in cell C2 and *copying that formula* throughout the indicated cells of Column C. Note that the formula is the *effective rate formula*, where **A2** is the annual rate and **B2** is m, the number of compoundings per year.

(a) Create Spreadsheet 6-3 on your computer.

(b) <u>Pencil and Paper Exercise</u>. Move down the effective rate column and compare the effective rates as m, the number of compoundings per year, increases. Verify that the effective rate corresponding to 5% compounded quarterly is 0.050945 or, equivalently, 5.0945%. State the effective rate corresponding to 5% compounded monthly. Can we conclude that for a constant annual rate, the effective rate increases as the number of compoundings per year increases?

(c) <u>Pencil and Paper Exercise</u>. Observe the effective rate corresponding to 5% compounded annually (i.e., $m = 1$). Can we conclude that with annual compounding, the effective rate equals the annual rate?

8. **Effective rate.** Create a spreadsheet similar to Spreadsheet 6-3 to compute the effective rates corresponding to 6% compounded annually, semiannually, quarterly, bimonthly, monthly and daily.

   (a) <u>Pencil and Paper Exercise</u>. State the effective rate corresponding to 6% compounded semiannually.

   (b) <u>Pencil and Paper Exercise</u>. State the effective rate corresponding to 6% compounded monthly.

   (c) <u>Pencil and Paper Exercise</u>. Move down the effective rate column and compare the effective rates as m, the number of compoundings per year, increases. Can we conclude that for a constant annual rate, the effective rate increases as the number of compoundings per year increases?

   (d) <u>Pencil and Paper Exercise</u>. Observe the effective rate corresponding to 6% compounded annually (i.e., $m = 1$). Can we conclude that with annual compounding, the effective rate equals the annual rate?

9. **Effective rate: Comparing two investments.** One investment pays 8.65% compounded quarterly while another pays 8.70% compounded semiannually. Assuming all else is the same, which is the better investment and why?

10. **Using GOAL SEEK to determine the annual rate.** Determine the annual rate needed to grow $10,000 into $15,000 during a time interval of 3 years. Assume monthly compounding.

   Substituting 10,000 for P, 12 for m, (3)(12) = 36 for n, and 15,000 for S into the compound amount formula

$$S = P\left(1 + \frac{r}{m}\right)^n$$

gives the equation

$$15000 = 10000\left(1 + \frac{r}{12}\right)^{36}$$

which we must solve for the annual rate, r, by using Excel's **Goal Seek** tool as illustrated in Spreadsheet 6-4 and the instructions that follow.

SPREADSHEET 6-4

r	S				
	10000				

INSTRUCTIONS FOR SPREADSHEET 6-4

1. Type the labels r and S as illustrated.

2. In cell B2, type the compound amount formula $=10000*(1+A2/12)^\wedge 36$, and press **Enter**. The result, 10000, appears in cell B2 as shown in Spreadsheet 6-4. Note that cell A2 represents the annual rate, r.

3. If it is not already there, use the mouse to *move the dark-bordered rectangle to the cell containing the formula*— in this case cell B2, which currently contains the value 10000.

4. Select **Tools**, then choose *Goal Seek* and a dialog box appears. Note that the *cell containing the formula*, in this case, cell B2, appears in the *Set cell* text box.

5. Because it is our goal to set cell B2 equal to a value of 15000 by changing the value of cell A2, we type **15000** in the *To value* text box, type **A2** in the *By changing cell* text box, and click **OK**. The required annual rate, r (in this case 0.135919), appears in cell A2. Thus, the required annual rate and solution to our problem is 13.5919% compounded monthly.

(a) Follow these instructions on your spreadsheet and verify the above results.
(b) Use your spreadsheet to recompute this problem, assuming the goal is to accumulate $25,000 instead of $15,000.

11. *Using GOAL SEEK to determine the annual rate.* Determine the annual rate needed to grow $40,000 into $180,000 during a time interval of 4 years. Assume quarterly compounding.

12. *Using GOAL SEEK to determine the annual rate.* Determine the annual rate needed to grow $30,000 into $100,000 during a time interval of 4 years. Assume annual compounding.

13. *Using GOAL SEEK to determine growth time.* Determine the time needed to grow $20,000 into $60,000 at an interest rate of 10% compounded semiannually.

14. *Using GOAL SEEK to determine growth time.* Determine the time needed to grow $15,000 into $90,000 at an interest rate of 18% compounded monthly.

15. *Using GOAL SEEK to determine growth time.* Determine the time needed to grow $18,000 into $120,000 at an interest rate of :
  (a) 10% compounded annually (b) 20% compounded annually (c) 25% compounded annually
  (d) *How the interest rate affects growth time*. Compare the results of parts (a) through (c). Does the growth time increase or decrease as the interest rate increases?

16. *Using GOAL SEEK to determine present value.* The principal needed to accumulate $90,000 during a time interval of 4 years at an interest rate of 30% compounded annually is called the *present value* of $90,000. Determining the present value involves solving the compound amount formula, $S = P(1+i)^n$, for P. With Excel, we can use *Goal Seek* to solve for P. Use your spreadsheet to find the present value for this problem.

17. *Using GOAL SEEK to determine present value.* Determine the principal needed to accumulate $90,000 during a time interval of 4 years at an interest rate of 10% compounded annually.

18. *Using GOAL SEEK to determine present value.* Determine the principal needed to accumulate $90,000 during a time interval of 5 years at an interest rate of 10% compounded quarterly.

19. *Using GOAL SEEK to determine present value: Zero-coupon bonds.* A zero-coupon bond is an investment contract that promises to pay its holder a fixed sum of money (called its *maturity value*) in the future. A typical problem faced by investors in zero-coupon bonds involves determining the price to pay for a zero-coupon bond having a given maturity value in order to earn a desired rate of return. This problem is equivalent to determining *present value* (or principal) of a future value.
   (a) Use *Goal Seek* to determine how much should be paid now for a 10-year zero-coupon bond having a maturity value of $100,000 if the investor wants to earn 15% compounded annually on this investment.
   (b) Use *Goal Seek* to determine how much should be paid now for a 10-year zero-coupon bond having a maturity value of $100,000 if the investor wants to earn 20% compounded annually on this investment.
   (c) Use *Goal Seek* to determine how much should be paid now for a 10-year zero-coupon bond having a maturity value of $100,000 if the investor wants to earn 30% compounded annually on this investment.

20. *How investors reap huge profits trading zero-coupon bonds.* Consider a 30-year zero-coupon bond having a maturity value of $100,000.
   (a) Use *Goal Seek* to determine how much should be paid now for this zero-coupon bond if the investor is to earn 7% compounded annually on this investment.
   (b) Suppose the investor bought this zero-coupon bond to earn 7% compounded annually and paid the price determined in part (a). One year later, market conditions have caused long-term interest rates to plummet to the point where this zero-coupon bond could be sold by the investor so that it provides its new holder an interest rate of only 6% compounded annually. Determine the price at which the current holder (investor) can sell this bond. Remember that only 29 years remain to maturity.
   (c) Suppose the original investor sold the bond for the price determined in part (b). Compare the price she or he paid [determined in part (a)] to the price she or he received [determined in part (b)] by computing the percent increase. Keep in mind that this percent increase was achieved during a 1-year time period. This illustrates how investors reap huge profits trading zero-coupon bonds. Profits depend upon declines in long-term interest rates.

6-2 International Investing

When investing internationally, one must consider currency exchange rates because currency fluctuations can either enhance or diminish the rate of return on a foreign investment. Specifically, if a foreign currency decreases in value, then the rates of return of foreign investments denominated in that currency will decrease, whereas if a foreign currency increases in value, the rates of return of foreign investments denominated in that currency will increase.

Spreadsheet 6-5 illustrates the example of $50,000 invested for 2 years in German bonds yielding 8% compounded annually. When this investment was made, the German mark (DM) was valued at $1.25, whereas when this investment matured, the German mark was valued at $1.10. The instructions following Spreadsheet 6-5 summarize the steps taken to determine the value of this investment at maturity in U.S. dollars.

SPREADSHEET 6-5

	PRESENT					FUTURE	
Initinvst $	50000						
Conv to fgn currency	40000				CompAmt	46656	FgnCurrcy
						51321.6	Conv to $

INSTRUCTIONS FOR SPREADSHEET 6-5

1. Type labels as needed.

2. Type **50000** in cell B3.

3. In cell B5, enter the formula **=B3/1.25** to convert $50,000 into German marks using the exchange rate $1.25 = 1 DM. The value 40,000 appears in cell B5.

4. In cell G5, enter the formula **=B5*1.08^2** to compute the compound amount of the 40,000 in cell B5.

5. In cell G7, enter the formula **=G5*1.10** to convert the compound amount of cell G5 into U.S. dollars using the exchange rate 1 DM = $1.10. The value 51,321.6 appears in cell G7, and this is the value of the investment at maturity in U.S. dollars, $51,321.60.

EXERCISES

1. *Determining compound amount.* Eighty thousand dollars is invested in Germany at 10% compounded annually for 3 years. When the investment was made, the German

mark was valued at $1.15. When the investment matured, the German mark was valued at $1.20. Use a spreadsheet to find the value of this investment at maturity in U.S. dollars.

2. *Determining compound amount.* Seventy thousand dollars is invested in Great Britain at 7% compounded annually for 4 years. When the investment was made, the British pound was valued at $1.50, whereas at maturity, the British pound was valued at $1.40. Use a spreadsheet to find the value of this investment at maturity in U.S. dollars.

3. *Determining compound amount.* Ninety thousand dollars is invested in Italy at 9% compounded annually for 2 years. When the investment was made, the Italian lira was valued at $0.000629, whereas at maturity, the lira was valued at $0.000610. Use a spreadsheet to find the value of this investment at maturity in U.S. dollars.

4. *Using GOAL SEEK to determine the annual rate.* Use Goal Seek to determine the annual rate of the investment in Exercise 1. Use the present and future values denominated in U.S. dollars.

5. *Using GOAL SEEK to determine the annual rate.* Use Goal Seek to determine the annual rate of the investment in Exercise 2. Use the present and future values denominated in U.S. dollars.

6. *Using GOAL SEEK to determine the annual rate.* Use Goal Seek to determine the annual rate of the investment in Exercise 3. Use the present and future values denominated in U.S. dollars.

7. *How currency appreciation can enhance a mediocre rate of return.* Ten thousand U.S. dollars is invested for 1 year in a stock whose value is denominated in Danish krones. The stock increased in value by only 2% during the year. When the investment was made, the U.S. dollar was valued at 5.7235 Danish krones, and 1 year later, the U.S. dollar had depreciated 30% against the Danish krone so that it was valued at $(0.70)(5.7235) = 4.0065$ krones. Note that the U.S. dollar depreciating against the krone means that the krone has appreciated against the dollar.
  (a) Use a spreadsheet to determine the maturity value of this investment in U.S. dollars.
  (b) Use Goal Seek to determine the annual rate of this investment. Use the present and future values denominated in U.S. dollars.
  (c) Note that this investment denominated in Danish krones increased by only 2%. However, the rate of return determined in part (b) is much larger because of the appreciation of the krone against the dollar.

6-3 Future Value of an Annuity

An annuity is a series of equal payments made at equal intervals of time. Specifically, a person investing $700 per quarter for 5 years, where each deposit earns interest at 6% compounded quarterly, is an example of an annuity. The *future value* of an annuity is defined as the *sum of the compound amounts of the individual payments*. Spreadsheet instructions for computing the future value of the above annuity are as follows.

INSTRUCTIONS

1. Move the dark-bordered rectangle to any cell and click f_x , *Paste Function*, on the toolbar. A dialog box appears. Use your mouse pointer to select *Financial* in the FUNCTION CATEGORY column and then select *FV* in the FUNCTION NAME column. Click *OK* at the bottom, and another dialog box appears.

2. Enter **6%/4** in the *rate* text box. Make certain that you include the percent sign (%). Note that because 6% is the annual rate, we must divide by 4 to get the corresponding quarterly rate.

3. Enter **5*4** or **20** in the *nper* text box. This gives the number of payment periods, n, of the annuity.

4. Enter **-700** in the *pmt* text box. This gives the periodic payment, R. Note that the payment is entered as a negative value. This results in a positive future value. Entering a positive payment would result in a negative future value.

5. Omit the *pv* and *type* text boxes. Note that omitting the *type* text box implies an ordinary annuity (payments made at the ends of periods), whereas entering **1** in the type text box implies an annuity due (payments made at the beginnings of periods).

6. Select **OK**, and the future value of the annuity appears in the cell with the dark-bordered rectangle.

Note: If ######## appears in a cell, this means that the number does not fit in the cell. One can widen a cell by moving the mouse pointer to the column heading at the top of the spreadsheet so that it points to the *right-hand border* of the *column*. The mouse pointer becomes a two-way arrow. *Double-click the right-hand border of the column*, and the number should appear in the cell. For the above example, the future value of the annuity, $16,186.57, appears in the cell with the dark-bordered rectangle.

EXERCISES

1. *Future value: Ordinary annuity.* Use a spreadsheet to compute the future value of each of the following ordinary annuities (payments made at the ends of periods).

(a) $600 quarterly for 6 years at 8% compounded quarterly.
(b) $900 monthly for 4 years at 7% compounded monthly.
(c) $500 semiannually for 9 years at 5% compounded semiannually.
(d) $2000 annually for 8 years at 6% compounded annually.

2. *Future value: Annuity due.* Use a spreadsheet to compute the future value of each of the annuities of Exercise, assuming each to be an annuity due (payments made at the beginnings of periods).

3. *Annuity due versus ordinary annuity.* Compare the results of Exercises 1 and 2 by dividing the future value of each annuity due by the future value of its corresponding ordinary annuity. Explain why the future value of an annuity due can be determined by multiplying the future value of its corresponding ordinary annuity by $(1 + i)$.

4. *Individual retirement account: Effect of annual rate on investment results.* A person invests $2000 at the end of each year into an individual retirement account (IRA) for 25 years. Find the future values corresponding to the interest rates (assume annual compounding) indicated in Spreadsheet 6-6. Compare the future values corresponding to interest rates of 4% and 20%.

SPREADSHEET 6-6

Rate	Nper	Pmt	FutValue				
4%	25	-2000					
8%	25	-2000					
12%	25	-2000					
16%	25	-2000					
20%	25	-2000					

INSTRUCTIONS FOR SPREADSHEET 6-6

1. Recreate Spreadsheet 6-6.

2. Move the dark-bordered rectangle to cell D2 in the FutValue column and use the instructions for computing the future value of an annuity. Once inside the *FV* dialog box, enter **A2** in the *rate* text box, **B2** in the *nper* text box, and **C2** in the *pmt* text box and select **OK**.

3. *Copy* the formula of cell D2 down through cell D6.

5. *Using GOAL SEEK to determine the annual rate.* Determine the annual rate needed for an ordinary annuity of $100 monthly to accumulate into $800,000 during a time interval of 20 years. Assume monthly compounding.

Solution: We begin by substituting 100 for R, 12 for m, $(20)(12) = 240$ for n, and 800,000 for S into the formula for the future value of an annuity

$$S = R\left[\frac{\left(1+\dfrac{r}{m}\right)^{n} - 1}{\dfrac{r}{m}}\right]$$

to obtain the equation

$$800000 = 100\left[\frac{\left(1+\dfrac{r}{12}\right)^{240} - 1}{\dfrac{r}{12}}\right]$$

which we must solve for the annual rate, r, by using Excel's **Goal Seek** tool as illustrated in Spreadsheet 6-7 and the instructions that follow.

SPREADSHEET 6-7

r	S				
0.06	46204.09				

INSTRUCTIONS FOR SPREADSHEET 6-7

1. Type the labels r and S as illustrated.

2. Type any annual rate, say 0.06, in cell A2.

3. In cell B2, type the formula for the future value of an annuity
=100*((1+A2/12)^240 - 1)/(A2/12) and press **Enter**. The result, 46204.09, appears in cell B2 as shown in Spreadsheet 6-7. Note that cell A2 represents the annual rate, r.

4. If it is not already there, use the mouse to *move the dark-bordered rectangle to the cell containing the formula*— in this case cell B2, which currently contains the value 46204.09.

5. Select **Tools**, and then choose **Goal Seek**. A dialog box appears. Note that the *cell containing the formula*, in this case cell B2, appears in the *Set cell* text box.

6. Because it is our goal to set cell B2 equal to a value of 800,000 by changing the value of cell A2, we type **800000** in the *To value* text box, type **A2** in the *By changing cell* text box, and click **OK**. The required annual rate, r (in this case 0.261047), appears in cell A2. Thus, the required annual rate and solution to our problem is 26.1047% compounded monthly.

(a) Follow these instructions on your spreadsheet and verify the above results.

(b) Use your spreadsheet to recompute this problem, assuming the goal is to accumulate $900,000 instead of $800,000.

6. *Using GOAL SEEK to determine the annual rate.* Determine the annual rate needed for an ordinary annuity of $500 per quarter to accumulate into $15,000 during a time interval of 4 years. Assume quarterly compounding.

7. *Using GOAL SEEK to determine the annual rate.* Determine the annual rate needed for an ordinary annuity of $2000 annually to accumulate into $17,000 during a time interval of 5 years. Assume annual compounding.

8. *Using GOAL SEEK to determine growth time.* Determine the time needed for an ordinary annuity of $1500 semiannually to accumulate into $60,000 at an interest rate of 10% compounded semiannually.

9. *Using GOAL SEEK to determine growth time.* Determine the time needed for an ordinary annuity of $300 per month to accumulate into $70,000 at an interest rate of 12% compounded monthly.

10. *Using GOAL SEEK to determine growth time.* Determine the time needed for an ordinary annuity of $1000 per year to accumulate into $16,000 at an interest rate of 10% compounded annually.

11. *Determining the payment R for a sinking fund.* A company foresees the need to replace piece of equipment 4 years from now at a projected cost of $50,000. The company wishes to accumulate $50,000 by setting aside monthly payments of R dollars each into a fund earning interest at 6% compounded monthly. Such a fund is called a *sinking fund*. The company must determine the monthly payment needed to accumulate the future value of $50,000.

Spreadsheet instructions for computing the *periodic payment* needed to accumulate the desired future value of the above annuity are as follows.

INSTRUCTIONS

1. Move the dark-bordered rectangle to any cell and click f_x , *Paste Function*, on the toolbar. A dialog box appears. Use your mouse pointer to select *Financial* in the FUNCTION CATEGORY column and then select *PMT* in the FUNCTION NAME column. Click **OK** at the bottom and another dialog box appears.

2. Enter 6%/12 in the *rate* text box. Make certain that you include the percent sign (%). Note that because 6% is the annual rate, we must divide by 12 to get the corresponding monthly rate.

3. Enter 4*12 or 48 in the *nper* text box. This gives the number of payment periods, n, of the annuity.

4. Enter **-50000** in the *fv* text box to enter the future value of the annuity. Note that the future value is entered as a negative value. This results in a positive payment. Entering a positive future value would result in a negative payment.

5. Omit the *pv* and *type* text boxes. Note that omitting the *type* text box implies an ordinary annuity (payments made at the ends of periods), whereas entering **1** in the type text box implies an annuity due (payments made at the beginnings of periods).

6. Select **OK**, and the periodic payment, $924.25, appears in the cell with the dark-bordered rectangle.

Note: If ######## appears in a cell, this means that the number does not fit in the cell. One can widen a cell by moving the mouse pointer to the column heading at the top of the spreadsheet so that it points to the *right-hand border* of the *column*. The mouse pointer becomes a two-way arrow. *Double-click the right-hand border of the column*, and the number should appear in the cell. For the above example, the future value of the annuity, $16,186.57, appears in the cell with the dark-bordered rectangle.

(a) Follow these instructions on your spreadsheet and verify the above results.
(b) Use your spreadsheet to recompute this problem, assuming the goal is to accumulate $90,000 instead of $50,000.

12. *Determining the payment R for a sinking fund.* A grocery chain anticipates the need for a new freezer 6 years from now when its cost is expected to be $60,000.
(a) If the grocery chain wishes to accumulate this amount by setting aside quarterly payments earning interest at 5% compounded quarterly for the next 6 years, how much should be set aside at the end of each quarter?
(b) Repeat part (a) assuming the goal is to accumulate $70,000 instead of $60,000.

13. *Determining the payment R for a sinking fund.* A contractor anticipates the need for a new truck 5 years from now when its cost is expected to be $40,000.
(a) If the contractor wishes to accumulate this amount by setting aside annual payments earning interest at 6% compounded annually for the next 5 years, how much should be set aside at the end of each year?
(b) Repeat part (a), assuming the goal is to accumulate $55,000 instead of $40,000.

14. *Determining the payment R for a sinking fund.* A person wishes to accumulate $250,000 in 20 years by setting aside annual payments earning interest at 5% compounded annually.
(a) How much should be set aside at the end of each year?
(b) Repeat part (a), assuming an interest rate of 15% compounded annually.
(c) Repeat part (a), assuming an interest rate of 25% compounded annually.
(d) *How the interest rate affects the payment.* Comparing the results of parts (a) through (c), does the payment, R, increase or decrease as the annual rate increases?

6-4 Present Value of an Annuity

The *present value* of an annuity is defined as the *sum of the present values of the individual payments*. Spreadsheet instructions for computing the present value of an annuity of $700 per quarter for 5 years at an interest rate of 6% compounded quarterly are as follows.

INSTRUCTIONS

1. Move the dark-bordered rectangle to any cell and click f_x , *Paste Function*, on the toolbar. A dialog box appears. Use your mouse pointer to select *Financial* in the FUNCTION CATEGORY column and then select *PV* in the FUNCTION NAME column. Click **OK** at the bottom, and another dialog box appears.

2. Enter **6%/4** in the *rate* text box. Make certain that you include the percent sign (%). Note that because 6% is the annual rate, we must divide by 4 to get the corresponding quarterly rate.

3. Enter **5*4** or **20** in the *nper* text box. This gives the number of payment periods, *n*, of the annuity.

4. Enter **-700** in the *pmt* text box. This gives the periodic payment, *R*. Note that the payment is entered as a negative value. This results in a positive present value. Entering a positive payment would result in a negative present value.

5. Omit the *fv* and *type* text boxes. Note that omitting the *type* text box implies an ordinary annuity (payments made at the ends of periods), whereas entering **1** in the type text box implies an annuity due (payments made at the beginnings of periods).

6. Select **OK** and the present value of the annuity appears in the cell with the dark-bordered rectangle.

Remember: If ######## appears in a cell, this means that the number does not fit in the cell. One can widen a cell by moving the mouse pointer to the column heading at the top of the spreadsheet so that it points to the *right-hand border* of the *column*. The mouse pointer becomes a two-way arrow. *Double-click the right-hand border of the column*, and the number should appear in the cell. For the above example, the present value of the annuity, $12,018.05, appears in the cell with the dark-bordered rectangle.

EXERCISES

1. *Present value: Ordinary annuity.* Use a spreadsheet to compute the future value of each of the following ordinary annuities (payments made at the end of periods).
  (a) $600 quarterly for 6 years at 8% compounded quarterly.

(b) $900 monthly for 4 years at 7% compounded monthly.

(c) $500 semiannually for 9 years at 5% compounded semiannually.

(d) $2000 annually for 8 years at 6% compounded annually.

2. *Present value: Annuity due.* Use a spreadsheet to compute the present value of each of the annuities of Exercise 1, assuming each to be an annuity due (payments made at the beginnings of periods).

3. *Annuity due versus ordinary annuity.* Compare the results of Exercises 1 and 2 by dividing the present value of each annuity due by the present value of its corresponding ordinary annuity. Explain why the present value of an annuity due can be determined by multiplying the present value of its corresponding ordinary annuity by $(1 + i)$.

4. *Interpretation of present value.* The *present value* of an annuity gives a *lump-sum value that, if invested now at the specified interest rate, will provide for the periodic payments of the annuity.* As an example, in Spreadsheet 6-8, we find the present value of an ordinary annuity of $1000 semiannually for 2 years at 8% compounded semiannually. Note that the present value, $3629.90, appears in cell E2 in the column labeled PVal A.

SPREADSHEET 6-8

AnnRte r	m	Term n	Pmt R	PVal A			
8%	2	4	-1000	$3,629.90			
Pmt #	Interest	Pmt	Balance				
0			$3,629.90				
1	145.20	-1000	$2,775.09				
2	111.00	-1000	$1,886.09				
3	75.44	-1000	$961.54				
4	38.46	-1000	$0.00				

Time Diagram for Spreadsheet 6-8

Balance:	$3629.90	$2775.09	$1886.09	$961.54	$0.00
		----------------	----------------	----------------	----------------	
Payments:		-$1000	-$1000	-$1000	-$1000

The Time Diagram for Spreadsheet 6-8 above illustrates the *meaning of present value of an annuity.* Specifically, $3629.90 invested *now* at 8% compounded semiannually allows one to withdraw $1000 semiannually (i.e., 4 payments) over 2 years. The computational details of this concept are revealed in the bottom portion of Spreadsheet 6-8. Note that the present value of the annuity, $3629.90, appears as the first entry corresponding to payment 0 in the Balance column. Studying the time diagram and the Payment 1 row of Spreadsheet 6-8, note that $145.20 is the interest earned on the balance of $3629.90 for a half-year. It was calculated by multiplying $3629.90 by the interest rate per period, 0.08/2

= 0.04. The new balance of $2775.09 was calculated by adding the interest of $145.20 to the previous balance, $3629.90, and then subtracting the $1000 payment (withdrawal). This process is repeated semiannually for 2 years (4 periods) to result in a final balance of $0.00. Instructions for Spreadsheet 6-8 are as follows. Note that the balances are also included in the time diagram.

INSTRUCTIONS FOR SPREADSHEET 6-8

1. Type labels as indicated in Spreadsheet 6-8.

2. Type **8%**, **2**, **4**, and **-1000** in cells A2, B2, C2, and D2, respectively.

3. Move the dark-bordered rectangle to cell E2 and click f_x on the toolbar to find the *present value of the annuity*.

4. For the bottom portion of the spreadsheet that reveals the meaning of present value of an annuity, type in **0**, **1**, and **2** in cells A6, A7, and A8, respectively. In cell D6, type the formula **=E2** to enter $3629.90.

5. In cell B7, type the formula **=$A$2*D6/$B$2** to calculate the interest earned on the balance, $3629.90, in cell D6. The dollar signs ($) denote an absolute reference to the respective cells, which means that the cell references do not change when copied to other cells. Also, if needed, choose **Format, Cells**, select **Number** in the Category box, and then click on the arrows in the Decimal places box until **2** appears in the box. Choose this combination of commands whenever two decimal places are needed.

6. In cell C7, type **=$D$2** to enter the $1000 payment in cell D2. Then copy this payment down to cell C8.

7. In cell D7, type the formula **=D6+B7+C7**, add the interest to the previous balance, and then subtract the withdrawal. Note that we use a plus sign before C7 because the minus sign is already included in cell C7.

8. In cell B8, type the formula **=$A$2*D7/$B$2** to calculate the interest earned on the balance, $2775.09, in cell D7.

9. In cell D8, type the formula **=D7+B8+C8** to add the interest to the previous balance, and then subtract the withdrawal.

10. This and the next step will copy the procedure of Steps 4 through 9 to complete the remaining schedule. Select cells A7 through D10 by moving the dark-bordered rectangle to cell A7. Then use the mouse to move the pointer inside cell A7, click the mouse button without releasing it, and drag the mouse pointer across to cell D7 and then down to cell D10. The selected area except for the initial cell (A7) will be in black.

11. Then choose the **Edit, Fill, Series** commands. In the dialog box that appears, make certain that Series in **Columns** is selected in the Type box. Next, select the **AutoFill** option and click **OK**. The complete schedule should appear on the spreadsheet.

(a) Create Spreadsheet 6-8.
(b) Return to Spreadsheet 6-8 created by using the above instructions. Move to cell A2 and change the annual rate to 7%. Note that the change is reflected throughout the spreadsheet. This is because we used cell references instead of constant numbers in Steps 4 through 9.
(c) Return to Spreadsheet 6-8 created by using the above instructions. Move to cell B2 and replace the 4 with **8** to change the term to 8 periods (i.e., 4 years instead of 2). Repeat Steps 10 and 11 to enlarge the schedule to 8 periods.

5. *Interpretation of present value.* Parents want to establish a fund for their daughter who is beginning college. The fund is to provide for monthly payments (end of month) of $300 for 4 years. Assume the money in this fund earns interest at 4% compounded monthly.
(a) Use a spreadsheet to determine the amount that should be invested in this fund now.
(b) Create a schedule (similar to that in Spreadsheet 4-8) that illustrates the interpretation of   present value.
(c) Use the results of part (b) to find the balance of this fund after the daughter makes her second withdrawal.
(d) Use the results of part (b) to find the balance of this fund after the daughter makes her twenty-fourth withdrawal.
(e) Pencil and Paper Exercise. Use a calculator to verify the interest earned by this fund just before the daughter makes her first withdrawal.
(f) Pencil and Paper Exercise. Use a calculator to verify the balance of this fund after the daughter makes her first withdrawal.
(g) Pencil and Paper Exercise. Use a calculator to verify the interest earned by this fund just before the daughter makes her second withdrawal.
(h) Pencil and Paper Exercise. Use a calculator to verify the balance of this fund after the daughter makes her second withdrawal.

6 – 9. *Another interpretation of present value.* Exercises 4 and 5 reveal a useful interpretation of present value of an annuity. The present value of an annuity also has the following equivalent interpretation. *The compound amount of the present value of an annuity equals the future value of the annuity.* Of course, this assumes the same interest rate and time periods for both annuity and compound amount. The following exercises illustrate this interpretation. For each of the following ordinary annuities:
(a) Use Excel to find the present value of the annuity.
(b) Use Excel to find the future value of the annuity.
(c) Use Excel to compute the compound amount of the answer to part (a).
(d) Compare the answers to parts (b) and (c). They should be equal.

6. An annuity of $600 per quarter for 8 years at 6% compounded quarterly.
7. An annuity of $800 semiannually for 7 years at 5% compounded semiannually..
8. An annuity of $1200 annually for 12 years at 9% compounded annually.

9. An annuity of $2500 per quarter for 6 years at 7% compounded quarterly.

10. ***Effect of the interest rate upon present value.*** Parents want to establish a fund for their child who is beginning college. The fund is to provide for monthly payments (end of month) of $500 for 4 years. Spreadsheet 6-9 illustrates a format to compare the effects of different rates upon the present value of the annuity.

SPREADSHEET 6-9

AnnRte r	m	Term n	Pmt R	PVal A			
2%	12	48	500				
3%	12	48	500				
4%	12	48	500				
5%	12	48	500				
6%	12	48	500				
7%	12	48	500				
8%	12	48	500				
9%	12	48	500				
10%	12	48	500				

(a) Create Spreadsheet 6-9 and fill in the present value (PVal A) column.
(b) Using the results of part (a), state how much money the parents should invest at an interest rate of 4% compounded monthly.
(c) Using the results of part (a), state how much money the parents should invest at an interest rate of 9% compounded monthly.
(d) Does the present value of the annuity increase or decrease as the interest rate increases?

6-5 Amortization

Amortization is the repayment of a loan by an annuity. When a loan is repaid by an annuity, it is said to be *amortized*. A portion of each payment is applied against the interest on the loan and the remainder is applied against the loan balance.

Determining the Loan Payment

When a loan is repaid by an annuity, *the amount of the loan is the present value of the annuity*. We use this concept to determine the monthly payment of a 3-year $7500 automobile loan at an interest rate of 7.9% compounded monthly. Spreadsheet instructions for computing the monthly payment are as follows.

INSTRUCTIONS

1. Move the dark-bordered rectangle to any cell and click f_x , *Paste Function*, on the toolbar. A dialog box appears. Use your mouse pointer to select *Financial* in the FUNCTION CATEGORY column and then select *PMT* in the FUNCTION NAME column. Click **OK** at the bottom, and another dialog box appears.

2. Enter **7.9%/12** in the *rate* text box. Make certain that you include the percent sign (%). Note that because 7.9% is the annual rate, we must divide by 12 to get the corresponding monthly rate.

3. Enter **3*12** or **36** in the *nper* text box. This gives the number of payment periods, n, of the annuity.

4. Enter **-7500** in the *PV* text box. This gives the amount of the loan (i.e., the present value of the annuity). Note that the present value is entered as a negative value. This results in a positive payment. Entering a positive present value would result in a negative payment.

5. Omit the *fv* and *type* text boxes. Note that omitting the *type* text box implies an ordinary annuity (payments made at the ends of periods), whereas entering **1** in the type text box implies an annuity due (payments made at the beginning of periods).

6. Select **OK**, and the payment, $234.68, appears in the cell with the dark-bordered rectangle.

Creating an Amortization Schedule

An amortization schedule shows the repayment of a loan on a payment-by-payment basis, where each payment is broken down into its interest and principal reduction (or balance reduction) components. As an illustrative example, Spreadsheet 6-10 illustrates an amortization schedule for the above $7500 aotumobile loan to be repaid by monthly payments over 3 years at an interest rate of 7.9% compounded monthly.

SPREADSHEET 6-10

LOAN PAYMENT CALCULATOR					Automobile Loan	
	AnnRate	Term n	LoanAmt	MoPmt	for	$7,500
	7.90%	36	7500	$234.68		
	AMORTIZATION SCHEDULE					
Pmt #	Payment	Interest	PrinRed	Balance		
0				7500		
1.00	234.68	49.38	185.30	7314.70		
2.00	234.68	48.16	186.52	7128.18		
3.00	234.68	46.93	187.75	6940.43		
4.00	234.68	45.69	188.99	6751.44		
5.00	234.68	44.45	190.23	6561.21		
6.00	234.68	43.19	191.48	6369.73		
7.00	234.68	41.93	192.74	6176.99		
8.00	234.68	40.67	194.01	5982.97		
9.00	234.68	39.39	195.29	5787.68		
10.00	234.68	38.10	196.57	5591.11		
11.00	234.68	36.81	197.87	5393.24		
12.00	234.68	35.51	199.17	5194.07		
13.00	234.68	34.19	200.48	4993.59		
14.00	234.68	32.87	201.80	4791.79		
15.00	234.68	31.55	203.13	4588.65		
16.00	234.68	30.21	204.47	4384.19		
17.00	234.68	28.86	205.81	4178.37		
18.00	234.68	27.51	207.17	3971.20		
19.00	234.68	26.14	208.53	3762.67		
20.00	234.68	24.77	209.91	3552.76		
21.00	234.68	23.39	211.29	3341.48		
22.00	234.68	22.00	212.68	3128.80		
23.00	234.68	20.60	214.08	2914.72		
24.00	234.68	19.19	215.49	2699.23		
25.00	234.68	17.77	216.91	2482.32		
26.00	234.68	16.34	218.33	2263.99		
27.00	234.68	14.90	219.77	2044.21		
28.00	234.68	13.46	221.22	1823.00		
29.00	234.68	12.00	222.68	1600.32		
30.00	234.68	10.54	224.14	1376.18		
31.00	234.68	9.06	225.62	1150.56		
32.00	234.68	7.57	227.10	923.46		
33.00	234.68	6.08	228.60	694.86		
34.00	234.68	4.57	230.10	464.76		
35.00	234.68	3.06	231.62	233.14		
36.00	234.68	1.53	233.14	0.00		

INSTRUCTIONS FOR SPREADSHEET 6-10

1. Determine the Monthly Payment
1.1 Type labels as indicated in Spreadsheet 6-10.

1.2 In cell B3 enter the annual interest rate, **7.9%**. Be sure to include the % sign.

1.3 In cell C3 enter **36**, the total number of payments.

1.4 In cell D3 enter **7500**, the amount of the loan.

1.5 Move the dark-bordered rectangle to cell E3 and click f_x , *Paste Function*, on the toolbar. A dialog box appears. Use your mouse pointer to select *Financial* in the FUNCTION CATEGORY column and then select *PMT* in the FUNCTION NAME column. Click **OK** at the bottom, and another dialog box appears.

1.6 Enter **B3/12** in the *rate* text box. This gives the monthly interest rate.

1.7 Enter **C3** in the *nper* text box. This gives the number of payment periods, *n*, of the annuity.

1.8 Enter **-D3** in the *pv* text box. This gives the loan amount, which is the present value of the annuity. Note that the loan amount is entered as a negative value. This results in a positive payment. Entering a positive loan amount would result in a negative payment.

1.9 Omit the *fv* and *type* text boxes. Note that omitting the *type* text box implies an ordinary annuity (payments made at the ends of periods), whereas entering **1** in the type text box implies an annuity due (payments made at the beginnings of periods).

1.10 Select **OK**, and the payment appears in the cell with the dark-bordered rectangle.

Remember: If **########** appears in a cell, this means that the number does not fit in the cell. One can widen a cell by moving the mouse pointer to the column heading at the top of the spreadsheet so that it points to the *right-hand border* of the *column*. The mouse pointer becomes a two-way arrow. *Double-click the right-hand border of the column*, and the number should appear in the cell. For the above example, the monthly payment, $234.68, appears in the cell with the dark-bordered rectangle.

2. Create the Amortization Schedule
2.1 Type labels as indicated in Spreadsheet 6-10.

2.2 Type the following entries in the indicated cells:

CELL	ENTRY
B6	**0**
F6	**=D3**
B7	**1**

B8	2
C7	=$E$3
C8	=$E$3

2.3 To determine the *interest* portion of the first payment, move to cell D7 and type the formula =**$B$3*F6/12**. The dollars signs ($) denote an absolute reference to the respective cells, which means that the cell references do not change when copied to other cells. Then choose **Format, Cells**, select **Number** in the Category box, and then click on the arrows in the Decimal places box until **2** appears in the box. Choose this combination of commands whenever two decimal places are needed.

2.4 To determine the *principal reduction* portion of the first payment, move to cell E7 and type the formula =**C7-D7**. Then move to cell F7 and type the formula =**F6-E7** to obtain the resulting *balance*.

2.5 To determine the *interest* portion of the second payment, move to cell D8 and type the formula =**$B$3*F7/12**. Again choose **Format, Cells**, select **Number** in the Category box, and then click on the arrows in the Decimal places box until **2** appears in the box.

2.6 To determine the *principal reduction* portion of the second payment, move to cell E8 and type the formula =**C8-D8**. Then move to cell F8 and type the formula =**F7-E8** to obtain the resulting *balance*. At this point, there should be an amortization schedule containing 2 payments.

2.7 To obtain the amortization schedule for all 36 payments:

First, select cells B7 through F42 by moving the dark-bordered rectangle to cell B7. Then use the mouse to move the pointer inside cell B7 until a white cross appears inside the cell, click the mouse button without releasing it, and drag the mouse pointer across to cell F7 and then down to cell F42. The selected area except for the initial cell (B7) will be in black.

Then choose the **Edit, Fill, Series** commands; make certain that Series in **Columns** is indicated; select the **AutoFill** option and click **OK**. The complete amortization schedule should appear on the spreadsheet.

EXERCISES

1. *Effect of interest rate upon monthly payment.* Construct a table giving the monthly payments for a 30-year mortgage for $100,000 at each of the following annual rates (compounded monthly): 8%, 7.5%, 7%, 6.5%, 6%. Use the format provided in Spreadsheet 6-11.

SPREADSHEET 6-11

AnnRate	m	Term n	LoanAmt	Payment			
8%	12	360	100000				
7.50%	12	360	100000				
7%	12	360	100000				
6.50%	12	360	100000				
6%	12	360	100000				

(a) Recreate Spreadsheet 6-11 and fill in the payment column.

(b) Analyzing the results of part (a), state whether the monthly mortgage payment increases or decreases as the annual rate decreases.

(c) Analyzing the results of part (a), each half-point reduction in the annual rate appears to decrease the monthly mortgage payment by approximately $_____.

2. *Effect of interest rate upon monthly payment*. Construct a table giving the monthly payments for a 15-year mortgage for $100,000 at each of the following annual rates (compounded monthly): 8%, 7.5%, 7%, 6.5%, 6%.

(a) Analyzing the above results, state whether the monthly mortgage payment increases or decreases as the annual rate decreases.

(b) Analyzing the above results, each half-point reduction in the annual rate appears to decrease the monthly mortgage payment by approximately $_____.

3. *Observations about 30-year mortgages.* Determine the monthly payment and amortization schedule for a 30-year $100,000 mortgage at an annual rate of 7% compounded monthly.

(a) *Note the mortgage balance after 20 years.* Determine the mortgage balance after: 10 years; 15 years; 20 years. Note that after 20 years, the mortgage balance is still greater than one-half the original mortgage. This is typical with 30-year mortgages.

(b) *Note the amount of interest paid during early years and later years.* Move down the amortization schedule, noting the *interest* and *principal reduction* components of each monthly payment. Note that for the early payments, most of the payment goes for interest and the remaining principal reduction is relatively small. However, for the later years, the situation reverses itself, with most of the payment going for principal reduction and a smaller portion going for the interest. However, note also that for this mortgage, the principal reduction component finally catches up to and slightly exceeds the interest component for the 231st monthly payment.

(c) *Note the time needed to repay one-half the original mortgage*. Continue to move down the amortization schedule and state the payment at which the mortgage balance is finally reduced to one-half the original mortgage. Notice that beyond this point, the mortgage balance finally begins to decrease substantially. Can we conclude that most of the principal reduction occurs during the last 10 years of this 30-year mortgage? This is typical of 30-year mortgages.

(d) Determine the total amount of interest paid on this mortgage over 30 years. Use the formula

Interest = sum of payments made - amount of loan.

4. *Observations about 30-year mortgages.* Repeat Exercise 3 for a 30-year $140,000 mortgage at an annual rate of 6.5% compounded monthly.

5. *Comparison: 15-year versus 30-year mortgages.* Determine the monthly payment and amortization schedule for a 15-year $100,000 mortgage at an annual rate of 7% compounded monthly.

(a) Compare this monthly payment to that of Exercise 3 by determining the percent increase of this monthly payment over that of Exercise 3.

(b) Determine the total amount of interest paid on this mortgage over 15 years. Compare this result to that of Exercise 3(d). Determine the amount of interest saved by using a 15-year mortgage instead of a 30-year mortgage.

(c) If one's budget allows for the larger monthly payment, does a 15-year mortgage appear to be a better deal for the borrower? State why or why not.

(d) Determine the mortgage balance after: 4 years; 7 years.

(e) Continue to move down the amortization schedule and state the payment at which the mortgage balance is finally reduced to one-half the original mortgage.

6. *Comparison: 15-year versus 30-year mortgages.* Repeat Exercise 5 for the mortgage in Exercise 4.

Mortgage Reduction Strategies

A *mortgage reduction strategy* is any increase in payment that results in a saving of interest paid by the borrower. As we learned in Exercise 5, a 15-year mortgage saves a substantial amount of interest for the borrower and is therefore one example of a mortgage reduction strategy. The following exercises illustrate other mortgage reduction strategies. Of course, a borrower should always check the terms of the mortgage to ensure that they allow for such additional payments (called *principal payments*) without penalties.

7. *Mortgage reduction: Making an additional payment.* Return to the amortization schedule of Exercise 3.

(a) Assume that an additional $1000 payment is made along with the 12th payment. Move the pointer to the cell.containing the 12th payment and retype 1665.30 in that cell.

(b) Move down the amortization schedule, observing the *balance* column, and note the payment at which the mortgage balance first becomes *negative*. In this example, this occurs at payment 349, where the balance after that payment is -28.83. This means that the 349th payment has overpaid the mortgage by $28.83.

(c) Determine the amount of interest saved (over the life of the mortgage) by making the additional $1000 payment.

8. *Mortgage reduction: Increasing the size of each payment.* Return to the amortization schedule of Exercise 3.

(a) Assume that an additional $10 payment is made along with each monthly payment. To accomplish this, use the following instructions.

114

INSTRUCTIONS

1. Change the first two monthly payments to 675.30. Then delete the remaining amortization schedule by moving the dark-bordered rectangle to the cell containing the third payment, use the mouse to move the pointer inside the cell, click the mouse button without releasing it, and drag the mouse pointer horizontally across to the last column of the amortization schedule and then down to the end of the amortization schedule. The *selected area* except for the initial cell will be in black.

2. Then choose **Edit, Delete, Entire Row** and click **OK.** The remaining amortization schedule should now be deleted.

3. To create the amortization schedule containing the increased monthly payment, move the dark-bordered rectangle to the cell containing the first monthly payment. Use the mouse to move the pointer inside the cell, click the mouse button without releasing it, and drag the mouse pointer horizontally across to the last column of the amortization schedule and then down to the end of the amortization schedule. Watch the row headings to ensure that the selected area provides for 360 monthly payments. The *selected area* except for the initial cell will be in black.

4. Then choose **Edit, Fill, Series** commands; select the **AutoFill** option and click **OK**. The complete amortization schedule should appear on your spreadsheet

(b) Move towards the end of the amortization schedule and determine the number of payments needed to repay the mortgage. Also, state the amount by which the last payment overpays the mortgage.
(c) Determine the amount of interest saved (over the life of the mortgage) by increasing the monthly payment.

9. *Mortgage reduction: Biweekly alternative.* One mortgage reduction strategy recently popularized is to convert a mortgage into a *biweekly* payment plan. This involves making a half-payment every 2 weeks, the net result being that an additional monthly payment is made during each year (52 weeks/2 equals 26 half-payments per year, or 13 monthly payments). Because a biweekly payment plan results in an additional monthly payment made during each 12 months, a biweekly payment plan can be *approximated* by increasing each monthly payment by one-twelfth of itself.

We illustrate such an approximation by returning to the amortization schedule of Exercise 3 and increasing each monthly payment by one-twelfth of itself. Specifically, begin by moving to the cell containing the first payment and typing the formula = **665.3/12+665.30**. Then modify the instructions of Exercise 8 to accomplish this mortgage reduction strategy.

Spreadsheet 6-12 shows the essential portions of the amortization schedule for the 30-year $100,000 mortgage at 7% (compounded monthly), and Spreadsheet 6-13 shows the essential portions of the same mortgage with the monthly payment increased by one-twelfth of itself.

SPREADSHEET 6-12

LOAN PAYMENT CALCULATOR						
	Interest	Term	Loan	Periodic		
	rate		amount	payment		
	7.00%	360	100000	665.3		
	AMORTIZATION SCHEDULE					
	Pmt#	Pmt	Int	PrinRed	Balance	
	0				100000	
	1	665.3	583.33	81.97	99,918.03	
	2	665.3	582.86	82.44	99.835.59	
	*	*	*	*	*	
	*	*	*	*	*	
	*	*	*	*	*	
	359	665.3	7.71	657.59	664.47	
	360	665.3	3.88	661.42	3.04	

In Spreadsheet 6-12, the balance of $3.04 after the last payment is made would normally be added to the last payment to give a resulting balance of $0.

SPREADSHEET 6-13

MONTHLY PAYMENT INCREASED BY ONE-TWELFTH OF ITSELF					
Pmt#	Pmt	Int	PrinRed	Balance	
0				100000	
1	720.74	583.33	137.41	99,862.59	
2	720.74	582.53	138.21	99,724.38	
*	*	*	*	*	
*	*	*	*	*	
284	720.74	8.09	712.65	674.04	
285	720.74	3.93	716.81	-42.77	

Studying Spreadsheet 6-13, observe that the mortgage is repaid with the 285th payment. In fact, the negative balance, -42.77, means that the 285th payment of $720.74 has overpaid the mortgage by $42.77. Thus the 285th payment would be decreased by $42.77, and the mortgage would be paid in full. In this case, increasing the monthly mortgae payment by one-twelfth of itself results in the mortgage being fully paid in 23.75 years, or approximately 6 years sooner than the corresponding 30-year mortgage.

Determining the Interest Saved
We begin by computing the *interest paid* on the 30-year mortgage with normal monthly payments as illustrated in Spreadsheet 6-12. The interest paid is determined by the formula

Interest = sum of payments made - amount of loan.

116

For the 30-year mortgage with *normal monthly payments*, the borrower will pay 360 payments of $665.30 for a total amount of (360)(665.30) = $239,508. Using the above formula, the interest paid is

$$\text{Interest} = \$239{,}508 - \$100{,}000 = \$139{,}508.$$

For the same mortgage with *monthly payments increased to $720.74* as illustrated in Spreadsheet 6-13, the borrower will make 284 payments of $720.74 each followed by a last payment of 720.74 - 42.77 = $677.97. Remember that, as we noted in our comments about Spreadsheet 6-13, the last payment was reduced by $42.77. Thus the borrower's total payments are

$$(284)(720.74) + 677.97 = \$205{,}368.13$$

and, therefore, the interest paid by the borrower is as follows:

$$\text{Interest} = \$205{,}368.13 - \$100{,}000 = \$105{,}368.13.$$

Comparing this against the interest paid using the normal monthly payment reveals a *savings* of

$$\$139{,}508 - \$105{,}368.13 = \$34{,}139.87$$

to the borrower. Note also that the mortgage is paid after 23.75 years.

10. *Mortgage reduction: Biweekly alternative.* Repeat Exercise 9 for the mortgage of Exercise 4.
 (a) Move towards the end of the amortization schedule and determine the number of payments needed to repay the mortgage. Also, state the amount by which the last payment overpays the mortgage.
 (b) Determine the amount of interest saved (over the life of the mortgage) by increasing the monthly payment.

11. *Mortgage reduction: Biweekly alternative.* Repeat Exercise 9 for a 30-year $80,000 mortgage at 6.75% compounded monthly.
 (a) Move towards the end of the amortization schedule and determine the number of payments needed to repay the mortgage. Also, state the amount by which the last payment overpays the mortgage.
 (b) Determine the amount of interest saved (over the life of the mortgage) by increasing the monthly payment.

6-6 Automobile Leases

In this section, we illustrate the mathematics-of-finance concepts used in automobile leasing. Specifically, we focus on the determination of the *monthly lease payment*, denoted by *R*, and the factors affecting its size. More precisely, the monthly payment on an auto lease depends upon the following four numbers.

1. The *term* of the lease. For example, if a 2-year lease is under consideration, then the lease term is 2 years.
2. The *capitalized cost*, denoted by *C*, of the car.
3. The *residual value*, *V*, of the car at the end of the lease term.
4. The *interest rate*, *i*.

We begin by discussing the *capitalized cost, C,* by considering the following example. Henry Jones signs a 3-year lease on a new car with a manufacturer's suggested retail price, abbreviated MSRP, of $25,000, a dealer discount of 10% of the MSRP, and a residual value that is 54% of the MSRP. Also, Henry is allowed a reduction in the capitalized cost of the new car by $1500 as trade-in value for his old car. Additionally, Henry further reduces the capitalized cost of the new car by making a $1000 down payment. The dealer, also charges Henry a $100 acquisition fee, which increases the capitalized cost. Finally, the dealer uses an interest rate of 7.5% compounded monthly. The determination of the capitalized cost is summarized as follows.

$25,000   MSRP
+ 100   Acquisition fee
- 2500   10% dealer discount (0.10 × 25,000 = $2500)
- 1500   Trade-in allowance for old car
- 1000   Deposit

$20,100   Capitalized cost

Note that future examples and exercises will not illustrate this breakdown. However, we present it here in the interest of providing a complete explanation.

The following time diagram illustrates the mathematics-of-finance concepts involved in the auto lease.

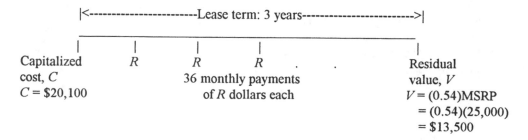

|<------------------------Lease term: 3 years------------------------>|

```
        |        |        |        |     .      .        |
Capitalized      R        R        R                    Residual
cost, C               36 monthly payments               value, V
C = $20,100              of R dollars each               V = (0.54)MSRP
                                                          = (0.54)(25,000)
                                                          = $13,500
```

These quantities are related by the formula

$$C = R\left[\frac{1-(1+i)^{-n}}{i}\right] + V(1+i)^{-n} \tag{1}$$

where $i = \dfrac{0.075}{12}$ and $n = (12)(3) = 36$. Note that the first component of the right-hand side of equation (1) is *the present value of the annuity of lease payments*, R, and the second component is the *present value of the residual value*, V. The present value of the residual value is also called the *discounted residual value*.

We solve equation (1) for R to determine the monthly lease payment as follows. Subtract $V(1+i)^{-n}$ from both sides to get

$$C - V(1+i)^{-n} = R\left[\frac{1-(1+i)^{-n}}{i}\right] \tag{2}$$

Now, the left-hand side of equation (2) is computed and treated as the present value of the annuity of monthly payments, R, for which we solve by using Excel's PMT function as illustrated in Spreadsheet 6-14.

SPREADSHEET 6-14

C	V(1 + i)^-n	C-V(1+i)^-n	PMT			
20100	10787.52	9312.480172	$289.68			

INSTRUCTIONS FOR SPREADSHEET 6-14

1. Type Labels and Formulas
1.1 After typing in the labels for each column, move to cell A2 and enter the *capitalized cost*, 20100.

1.2 Move to cell B2 and enter the formula =13500*(1+.075/12)^(-36) for the *discounted residual value*.

1.3 Move to cell C2 and enter the formula =A2-B2 to compute the left-hand side of equation (2). This result is treated as the *present value of the annuity of monthly lease payments*.

2. Compute the monthly lease payment by using Excel's PMT function

2.1 Move the dark-bordered rectangle to cell D2 and click f_x , **Paste Function**, on the toolbar. A dialog box appears. Use your mouse pointer to select **Financial** in the FUNCTION CATEGORY column and then select **PMT** in the FUNCTION NAME column. Click **OK** at the bottom, and another dialog box appears.

2.2 Enter **7.5%/12** in the *rate* text box. Make certain that you include the percent sign (%) or, instead, enter **0.075/12**. Note that because 7.5% is the annual rate, we must divide by 12 to get the corresponding monthly rate.

2.3 Enter **3*12** or **36** in the *nper* text box. This gives the number of payment periods, n, of the annuity.

2.4 Enter **-C2** in the *PV* text box. Remember, the content of cell C2 is treated as the present value of the annuity of monthly lease payments. Note that the present value is entered as a negative value. This results in a positive payment. Entering a positive present value would result in a negative payment.

2.5 Omit the *fv* and *type* text boxes. Remember, omitting the *type* text box implies an ordinary annuity (payments made at the ends of periods), whereas entering **1** in the type text box implies an annuity due (payments made at the beginnings of periods).

2.6 Select **OK**, and the payment, $289.68, appears in the cell D2. Thus, the monthly lease payment is $289.68.

EXERCISES

1. *Determining the monthly lease payment.* Use the format of Spreadsheet 6-14 to determine the monthly lease payment for each of the following automobile leases. Note that all interest rates assume monthly compounding.
   (a) A 2-year lease with a capitalized cost of $18,000, a residual value of $6000, and an annual interest rate of 6.5%.
   (b) A 3-year lease with a capitalized cost of $15,000, a residual value of $4500, and an annual interest rate of 7%.
   (c) A 2-year lease with a capitalized cost of $25,000, a residual value of $8000, and an annual interest rate of 6.75%.

2. *Effect of capitalized cost upon monthly payment.* Construct a table giving the monthly payments for the 3-year automobile lease of Spreadsheet 6-14, assuming different capitalized costs as shown in Spreadsheet 6-15. Recall that the interest rate is 7.5% compounded monthly and the residual value is $13,500.

SPREADSHEET 6-15

C	V(1+i)^-n	C-V(1+i)^-n	PMT			
20100	10787.52	9312.48	$289.68			
20600	10787.52					
21100	10787.52					
21600	10787.52					
22100	10787.52					
22600	10787.52					
23100	10787.52					

(a) Complete the two unfilled columns of Spreadsheet 6-15.

(b) Pencil and Paper Exercise. State what happens to the monthly lease payment as the capitalized cost increases.

(c) Pencil and Paper Exercise. Is there a large increase in the monthly lease payment between the original capitalized cost and the largest capitalized cost given in the spreadsheet?

3. *Effect of capitalized cost upon monthly payment.* Repeat Exercise 2 for a 2-year automobile lease with capitalized costs given in Spreadsheet 6-16, a residual value of $5000, and an interest rate of 6.75% compounded monthly.

SPREADSHEET 6-16

C	V(1+i)^-n	C-V(1+i)^-n	PMT			
15000						
15500						
16000						
16500						
17000						
17500						

4. *Effect of residual value upon monthly payment.* Construct a table giving the monthly payments for the 3-year automobile lease of Spreadsheet 6-14, assuming different residual values as shown in Spreadsheet 6-17. Recall that the capitalized cost is $20,100, the interest rate is 7.5% compounded monthly, and the original residual value is $13,500.

C	Residual, V	$(1+i)^{-n}$	$V(1+i)^{-n}$	$C-V(1+i)^{-n}$	PMT	
20100	13500					
20100	14000					
20100	14500					
20100	15000					
20100	15500					
20100	16000					
20100	16500					

(a) Complete the four unfilled columns of Spreadsheet 6-17.

(b) Pencil and Paper Exercise. State what happens to the monthly lease payment as the residual value increases.

(c) Pencil and Paper Exercise. Is there a large increase in the monthly lease payment between the original residual value and the largest residual value given in the spreadsheet?

5. *Difference between capitalized cost and discounted residual value: A summary.*

(a) Pencil and Paper Exercise. Explain why decreasing the quantity $C-V(1+i)^{-n}$ results in a lower monthly lease payment.

(b) Pencil and Paper Exercise. Explain why decreasing the capitalized cost while keeping the residual value at a constant level will decrease the monthly lease payment.

(c) Pencil and Paper Exercise. Explain why increasing the residual value while keeping the capitalized cost at a constant level will decrease the monthly lease payment.

Summary. Parts (a) through (c) have revealed that the smaller the difference between the capitalized cost and residual value, the smaller will be the monthly lease payment. This makes sense intuitively because, when one leases a car, one is paying for the car's depreciation during the term of the lease. Depreciation is defined as the difference between the capitalized cost and the residual value. Thus, the smaller the depreciation, the smaller the monthly lease payment.

6. *Effect of depreciation on the monthly lease payment.* Construct a table giving the monthly payments for a 2-year automobile lease, assuming different capitalized costs and residual values as shown in Spreadsheet 6-18. Assume the interest rate is 6.5% compounded monthly. Also note that as we move down the columns of Spreadsheet 6-18, the capitalized costs are decreasing while the residual values are increasing. Thus, the depreciation is decreasing as we move down the columns.

SPREADSHEET 6-18

C	V	(1+i)^-n	V(1+i)^-n	C-V(1+i)^-n	PMT	
16000	5000					
15800	5200					
15600	5400					
15400	5600					
15200	5800					
15000	6000					
14800	6200					
14600	6400					
14400	6600					

(a) Complete the four unfilled columns of Spreadsheet 6-18.

(b) Pencil and Paper Exercise. State what happens to the monthly lease payment as the depreciation decreases.

(c) Pencil and Paper Exercise. State the decrease in the monthly lease payment between the largest and smallest depreciation values given in the spreadsheet.

7. *Decreasing the capitalized cost.* Return to the first page of this section, where the components of capitalized cost are listed.

(a) Pencil and Paper Exercise. State the effect of a decrease in the MSRP on capitalized cost.

(b) Pencil and Paper Exercise. State the effect of an increase in the trade-in allowance for one's old car on capitalized cost.

(c) Pencil and Paper Exercise. State the effect of an increase in the deposit on capitalized cost.

CHAPTER SEVEN

Probability Distributions

7-1 Probability Distributions and Expected Value

Frequently, in business and other disciplines, chance events take the form of numerical outcomes. The numerical outcomes are denoted by a letter—say, x—which is called a *random variable*. As an example, consider a simple type of gambling machine, a box full of numbers as illustrated in Spreadsheet 7-1. Observing the gambling machine in Spreadsheet 7-1, note that the box contains the numbers 1, 3, 5, 8, and 10. Specifically, in the proportions shown, 20% of all the numbers in the box are 1s, 50% are 3s, 15% are 5s, 10% are 8s, and 5% are 10s.

A chance experiment consists of selecting 1 number from the box. The number selected will be taken to a cashier who will pay the player a dollar amount equal to the number. The percentages in Spreadsheet 7-1 indicate that the player has a 20% probability of winning $1, a 50% probability of winning $3, a 15% probability of winning $5, a 10% probability of winning $8, and a 5% probability of winning $10.

SPREADSHEET 7-1

GAMBLING MACHINE				PROBABILITY DISTRIBUTION		
Box full of numbers					x	P(x)
1	20%				1	0.20
3	50%				3	0.50
5	15%				5	0.15
8	10%				8	0.10
10	5%				10	0.05
	100%					1.00

The table in the upper right-hand corner of Spreadsheet 7-1 gives the numerical outcomes of this chance experiment listed alongside their corresponding probabilities. The numerical outcomes are denoted by a letter, X, and the probabilities are denoted by $P(X)$. Because the values of X are chance outcomes, X is called a *random variable*. Such a display of values of a random variable and their corresponding probabilities is called a *probability distribution* as indicated in Spreadsheet 7-1. *Note that the probabilities sum to 1.*

Expected Value

Given a random variable X that assumes the values $x_1, x_2, \ldots, x_n$ with corresponding probabilities $p_1, p_2, \ldots, p_n$, respectively, the *expected value* of X is denoted by *E(X)* and is determined as follows:

$$E(X) = x_1 p_1 + x_2 p_2 + \cdots + x_n p_n$$

Spreadsheet 7-2 illustrates the calculation of the expected value, $E(X)$, for the gambling machine.

SPREADSHEET 7-2

X	P(X)	XP(X)	GAMBLING MACHINE		
1	0.2	0.20			
3	0.5	1.50	Interpretation of Expected Value		
5	0.15	0.75	If one were to play this gambling		
8	0.1	0.80	machine over and over again, one		
10	0.05	0.50	would win, on the average, $3.75		
	1	3.75 ←——E(X	per play.		

The expected value is a weighted average (where the probabilities are the weights) of the values of a random variable. Notice the interpretation of the expected value given in Spreadsheet 7-2. We add to this interpretation by noting that if one were to play this gambling machine over and over again, one would win $1 20% of the time, $3 50% of the time, $5 15% of the time, $8 10% of the time, and $10 5% of the time. The result is a *long-run average* of $3.75 per play.

INSTRUCTIONS

Use the following to calculate the expected value as illustrated in Spreadsheet 7-2.

1. Type the label **X** in cell A1, **P(X)** in cell B1, and **XP(X)** in cell C1 as illustrated in Spreadsheet 7-2.

2. Type the values of the random variable X in cells A2 through A6.

3. Type the probability values in cells B2 through B6.

4. Move to cell C2 and type the formula **=A2*B2** and press **Enter**. This gives the product, 0.20, of the contents of cells A2 and B2.

5. *Copy this formula* down through cell C6. This gives the products of the values of the random variable and their corresponding probabilities.

6. *Sum* the products of cells C2 through C6. Move the mouse pointer to the middle of cell C2, click, hold, and drag the mouse pointer down to cell C6. Cells C3 through C6 should be highlighted with cell C2 remaining unhighlighted. Click $\sum$ on the toolbar, and the sum appears in cell C7. Move the mouse pointer to cell C7 and click on **boldface** to distinguish the **sum** from the column entries.

EXERCISES

1. *Probability distribution: Missing probabilities.* The following spreadsheet gives three probability distributions. Fill in the missing probability for each distribution.

X	P(X)		X	P(X)		X	P(X)
1	0.10		1	0.05		2	0.13
3	0.30		2	0.15		5	0.27
5			3	0.30		9	0.30
9	0.20		4	0.27		13	
12	0.15		5			19	0.18
	1			**1**			**1**

2. *Comparing distributions: Expected value.* The following spreadsheet gives probability distributions of dollar value outcomes for two gambling machines.

Gambling Machine 1				Gambling Machine 2		
	X	P(X)			X	P(X)
	1	0.30			1	0.02
	2	0.50			2	0.08
	3	0.10			3	0.10
	4	0.08			4	0.50
	5	0.02			5	0.30
		1				**1**

(a) Use a spreadsheet to determine the expected value for each gambling machine.
(b) <u>Pencil and Paper Exercise</u>. Explain why the expected value for Gambling Machine 2 is much greater than the expected value for Gambling Machine 1. Hint: Compare the distributions of probabilities for these machines.
(c) <u>Pencil and Paper Exercise</u>. If the casino wants to make an expected profit of $1 per play for Gambling Machine 1, determine the amount it must charge per play.
(d) <u>Pencil and Paper Exercise</u>. If the casino charges gamblers $6 per play for Gambling Machine 2, determine the expected profit per play for Gambling Machine 2.

3. *Insurance.* An insurance company sells a $20,000 life insurance policy. The insured must pay an annual premium of $350. If actuarial tables reveal that the probability of death of the insured during the next year is .002, then the probability distribution of dollar gain to the insurance company is given in the following spreadsheet.

Outcomes	Gain to insurance company X		P(X)		
Insured dies	$350 - $20,000 = ——>	-19650	0.002		
Insured lives	$350 ————>	350	0.998 <————	1 - 0.002	
			1		

(a) Use a spreadsheet to compute the expected value (gain) to the insurance company.

(b) Pencil and Paper Exercise. Assume that the expected value computed in part (a) is $310. Does this mean that the insurance company will make a profit of $310 if it sells only one policy? Explain.

4. *Insurance.* An insurance company sells a $30,000 life insurance policy. The insured must pay an annual premium of $510. Assume that actuarial tables reveal that the probability of death of the insured during the next year is .001.

(a) Use a spreadsheet to determine the probability distribution of dollar gain to the insurance company.

(b) Use a spreadsheet to compute the expected value (gain) to the insurance company.

(c) Pencil and Paper Exercise. Does the expected value determined in part (b) give the insurance company's profit from selling only one policy? Explain.

5. *Daily demand: Converting frequencies to probabilities.* The manager of a hamburger stand has kept a record of daily demand for hamburgers during the past 400 days. The results are shown in the following spreadsheet.

X	f		P(X)		
(Demand)	(Number of days)				
100	20			<————	20/400
110	60			<————	60/400
120	200			<————	200/400
130	80			<————	80/400
140	40			<————	40/400
	400				

(a) Use a spreadsheet to convert the frequency distribution into a probability distribution by computing the percent frequency of occurrence for each demand level as indicated in the spreadsheet.

(b) Use a spreadsheet to compute the expected demand.

(c) Pencil and Paper Exercise. Interpret the expected value found in part (b).

6. *Daily demand: Converting frequencies to probabilities.* The manager of a pizza stand has kept a record of weekly demand for pizzas during the past 50 weeks. The results are shown in the following spreadsheet.

X		f
(Demand)	(Number of weeks)	
20000		5
25000		10
30000		25
35000		5
40000		5
		50

(a) Use a spreadsheet to convert the frequency distribution into a probability distribution by computing the percent frequency of occurrence for each demand level.

(b) Use a spreadsheet to compute the expected demand.

(c) Pencil and Paper Exercise. Interpret the expected value found in part (b).

7-2 Variance and Standard Deviation

The **variance** of a random variable indicates the degree of spread or dispersion of the values of the random variable about the expected value (or mean). Given a random variable X that assumes the values $x_1, x_2, \ldots, x_n$ with corresponding probabilities $p_1, p_2, \ldots, p_n$, respectively, and an expected value of μ, the **variance** of X is denoted by Var(X) and is determined as follows:

$$Var(X) = p_1(x_1 - \mu)^2 + p_2(x_2 - \mu)^2 + \ldots + p_n(x_n - \mu)^2$$

The **standard deviation** of a random variable is denoted by σ (pronounced "sigma") and is the *square root of the variance*.

Spreadsheet 7-3 illustrates the computation of the variance for two probability distributions.

SPREADSHEET 7-3

X	P(X)	XP(X)	X - u	(X - u)^2	(X - u)^2P(X)	
1	0.1	0.1	-2	4	0.4	
2	0.2	0.4	-1	1	0.2	
3	0.4	1.2	0	0	0	
4	0.2	0.8	1	1	0.2	
5	0.1	0.5	2	4	0.4	
	1	3 <—Expected value			1.2 <——Variance	
X	P(X)	XP(X)	X - u	(X - u)^2	(X - u)^2P(X)	
1	0.05	0.05	-2	4	0.2	
2	0.1	0.2	-1	1	0.1	
3	0.7	2.1	0	0	0	
4	0.1	0.4	1	1	0.1	
5	0.05	0.25	2	4	0.2	
		3 <—Expected value			0.6 <—Variance	

Comparing both distributions in Spreadsheet 7-3, observe that the random variable takes on the same values in both distributions. However, observe that the probabilities (or, equivalently, the weights) for the bottom distribution are more concentrated near the center (or mean) of the distribution and therefore less spread out than those for the top distribution. Note that for the top distribution, larger probabilities are assigned to values away from the mean than for the bottom distribution. Accordingly, the variance of the top distribution, 1.2, is larger than the variance of the bottom distribution, 0.6.

Additional insight into the meaning of variability is gained by observing the computations of the variance for both distributions. Specifically, note that the deviations **X - u** are the same for both distributions but the **(X - u)^2P(X)** values differ because of the redistribution of the probabilities (or weights). Notice that for the top distribution, the large contributions, 0.4, are made to the $(X - u)^2 P(X)$ column from the x-values, 1 and 5, farthest away from the center (or mean). This is an additional result of the top distribution having greater variability.

INSTRUCTIONS

Use the following steps to perform the calculations illustrated for the top distribution in Spreadsheet 7-3.

1. Type labels as required.
2. Type the values of X in cells A2 through A6 and values of $P(X)$ in cells B2 through B6 as indicated.
3. Compute the expected value by moving to cell C2 and typing the formula **=A2*B2**. *Copy* this formula down through cell C6 and click $\sum$ on the toolbar to obtain a *sum* of the products in cell C7.
4. Calculate the deviations by moving to cell D2, typing the formula **=D2-$C$7,** and *copying* the formula down through cell D6. Note that the dollar signs ($) indicate an absolute cell reference to cell C7 which means that this cell does not change when the formula is copied down through cell D6.
5. Calculate the squares of the deviations by moving to cell E2 and typing the formula **=D2^2**. *Copy* the formula down through cell E6.
6. Multiply the squares of the deviations by their respective probabilities and sum the results by moving to cell F2, typing the formula **=E2*B2**, *copying* the formula down through cell F6, and clicking on $\sum$ to obtain a *sum* of the results in cell F7.

EXERCISES

1. *Comparing probability distributions: Variability*. Consider the two distributions in the following spreadsheet.

X	P(X)			X	P(X)		
1	0.05			1	0.15		
2	0.15			2	0.20		
3	0.60			3	0.30		
4	0.15			4	0.20		
5	0.05			5	0.15		
	1.00				1.00		

(a) <u>Pencil and Paper Exercise</u>. State which distribution appears to have greater variability and explain why.
(b) Use a spreadsheet to calculate the variance for each distribution.

(c) Use the spreadsheet to calculate the standard deviation for each distribution.

(d) <u>Pencil and Paper Exercise</u>. Compare the entries in the *(X - u)^2P(X)* column for both distributions. For the distribution having the greater variance, note the *x*-values resulting in large contributions to the variance.

2. *Comparing probability distributions: Variability*. Consider the two distributions in the following spreadsheet.

X	P(X)			X	P(X)	
1	0.05			1	0.30	
2	0.20			2	0.50	
3	0.50			3	0.10	
4	0.20			4	0.05	
5	0.05			5	0.05	
	1.00				1.00	

(a) <u>Pencil and Paper Exercise</u>. State which distribution appears to have greater variability and explain why.

(b) Use a spreadsheet to calculate the variance for each distribution.

(c) Use the spreadsheet to calculate the standard deviation for each distribution.

(d) <u>Pencil and Paper Exercise</u>. Compare the entries in the **(X - u)^2P(X)** column for both distributions. For the distribution having the greater variance, note the *x*-values resulting in large contributions to the variance.

3. *Overcontrol of a production process.* The following spreadsheet gives probability distributions of diameters of a circular disk used inside telecommunications networking equipment. The production equipment is calibrated to produce disks having diameters of 2.00 cm. The distribution labeled *Normal Control* results from minimal interference by operators into the production process, whereas the distribution labeled *Overcontrol* results from an operator continually modifying process controls in an attempt to keep the disk diameters near the center (or mean).

Normal Control				Overcontrol		
X	P(X)			X	P(X)	
1.96	0.01			1.96	0.05	
1.97	0.03			1.97	0.15	
1.98	0.09			1.98	0.14	
1.99	0.12			1.99	0.11	
2.00	0.50			2.00	0.10	
2.01	0.12			2.01	0.11	
2.02	0.09			2.02	0.14	
2.03	0.03			2.03	0.15	
2.04	0.01			2.04	0.05	

131

(a) <u>Pencil and Paper Exercise</u>. Comment on the distribution of the probabilities in both cases. Specifically, for each distribution, state where the heavier concentration of probabilities occurs and how that differs from the other distribution.

(b) Use a spreadsheet to calculate the variance for each distribution.

(c) Use the spreadsheet to calculate the standard deviation for each distribution.

(d) <u>Pencil and Paper Exercise</u>. Compare the entries in the $(X - u)\wedge 2P(X)$ column for both distributions. For the distribution having the greater variance, note the x-values resulting in large contributions to the variance.

(e) <u>Pencil and Paper Exercise</u>. State which distribution has greater variability.

(f) <u>Pencil and Paper Exercise</u>. Textbooks in statistical process control teach that overcontrol of a production process leads to increased variability of the product produced. State your insights as to what's happening to the overcontrolled process in the above spreadsheet and why the probabilities are distributed as indicated.

4. *Fear distribution.* A producer of manufacturing equipment for the telecommunications industry buys metal rods from a supplier. The rods are designed so that their lengths satisfy specification limits of 3.00 cm $\pm$ 0.02 cm. The following spreadsheet gives a distribution of rod lengths from a production lot recently purchased from the supplier.

	X	P(X)				
	2.96	0.04				
	2.97	0.00				
	2.98	0.20				
	2.99	0.17				
	3.00	0.30				
	3.01	0.17				
	3.02	0.10				
	3.03	0.02				

(a) <u>Pencil and Paper Exercise</u>. Note the zero probability corresponding to the x-value 2.97, which is just below the lower specification limit of 2.98 cm. What does this suggest?

(b) Use a spreadsheet to compute the variance.

7-3 The Binomial Distribution

The binomial distribution gives probabilities of getting x successes out of n independent trials, where p denotes the probability of getting a success on an individual trial. Individual probabilities are computed in accordance with the formula

$$C(n,x)p^x(1-p)^{n-x}$$

Here $C(n,x)$ is the number of ways of selecting x things out of a set of n things where the order of selection does not matter. Spreadsheet 7-4 gives a binomial distribution for $n = 5$ and $p = 0.2$

SPREADSHEET 7-4

Binomial Distribution		$n = 5$, $p = 0.2$					
X	P(X)						
0	0.32768						
1	0.4096						
2	0.2048						
3	0.0512						
4	0.0064						
5	0.00032						
	1						

INSTRUCTIONS

Use the following steps to obtain the binomial distribution in Spreadsheet 7-4.

1. Type labels as required.
2. Type the values of X in cells A3 through A8 as indicated.
3. Move the dark-bordered rectangle to cell B3. Click f_x , **Paste Function**, on the toolbar. A dialog box appears. Use your mouse pointer to select **Statistical** in the FUNCTION CATEGORY column and then select **Binomial** in the FUNCTION NAME column. Click **OK**, and another dialog box appears.
4. Enter **A3**, the number of successes (x) in the **Number** box. We would enter 0 if our wish were to obtain the probability for 0 successes only. However, because it is our intention to copy this formula for the indicated x-values, we enter A3.
5. Enter **5**, the number of trials (n) in the **Trials** box.
6. Enter **.2**, the probability (p) of success per trial in the **Probability** box.
7. Enter **0** in the **Cumulative** box to obtain the individual probability. Entering 1 here will return the cumulative probability.
8. Click **OK** and the probability appears in cell B3 with the dark-bordered rectangle.
9. *Copy* the formula down through cell B8.

Expected Value and Variance

The expected value (or mean) and variance of a binomial distribution are given by the following formulas.

Expected Value: $E(X) = np$ Variance: $Var(X) = np(1 - p)$

EXERCISES

1. *Comparing distributions.* Use a spreadsheet to obtain the binomial distributrion for:
   (a) $n = 8$ and $p = 0.1$ (b) $n = 8$ and $p = 0.5$ (c) $n = 8$ and $p = 0.9$
   (d) Study the distribution of probabilities for part (a). State whether the larger probabilities occur at smaller or larger values of X. Explain why.
   (e) Study the distribution of probabilities for part (c). State whether the larger probabilities occur at smaller or larger values of X. Explain why.
   (f) Study the distribution of probabilities for part (b). State whether the larger probabilities occur at smaller or larger values of X. Explain why.
   (g) Compare the probabilities for parts (a) and (c). Explain why both distributions have the same probabilities, but in reverse order.

2. *Quality control.* A production lot of small electrical components contains 10% defectives. A quality control procedure consists of selecting 5 components from the lot. Despite the fact that there is no replacement between selections, we will assume independence because there are so many components in the production lot that the proportion of defectives (10%) remains nearly constant as the selections begin.
   (a) Use a spreadsheet to obtain the binomial distribution for the number of defectives obtained out of the 5 selections.
   (b) State the probability of getting exactly 0 defectives out of the 5 selections.
   (c) State the probability of getting exactly 2 defectives out of the 5 selections.
   (d) Compute the expected value, variance, and standard deviation for this binomial distribution.

3. *Medical testing.* A new drug for treating a given disease has been proven to be effective 70% of the time. Assume the drug is administered to 9 patients.
   (a) Use a spreadsheet to obtain the binomial distribution for the number of cures out of the 9 patients.
   (b) State the probability of getting exactly 5 cures out of the 9 patients.
   (c) State the probability of getting exactly 7 cures out of the 9 patients.
   (d) Compute the expected value, variance, and standard deviation for this binomial distribution.

4. *Sales presentation.* Each time a particular salesperson makes a presentation, there is a 20% chance that the customer will buy the product. The salesperson usually makes 5 presentations per day. Let x represent the number of customers who buy the product during an ordinary day.
   (a) Use a spreadsheet to obtain the binomial distribution for the number of customers who buy the product during an ordinary day.

134

(b) State the probability that exactly 2 customers buy the product during an ordinary day.
(c) State the probability that exactly 3 customers buy the product during an ordinary day.
(d) On the average, this salesperson should expect to make how many sales per day?
(e) Use a spreadsheet to calculate the variance and standard deviation.

5. *Advertising effectiveness.* An advertising agency claims that 60% of the viewers of a particular television commercial will buy the product. A chance experiment consists of surveying 10 viewers of this commercial. Let x denote the number of viewers who bought the product.
(a) Use a spreadsheet to obtain the binomial distribution for x, the number of viewers who bought the product.
(b) Use a spreadsheet to compute the expected value, variance, and standard deviation.
(c) Pencil and Paper Exercise. Suppose only 2 out of the 10 viewers bought the product. What might we conclude about the agency's claim that $p = 0.6$?

6. *Binomial distribution: The effect of p upon the variance.* Consider the following spreadsheet.

n	p	Variance				
10	0.1					
10	0.2					
10	0.3					
10	0.4					
10	0.5					
10	0.6					
10	0.7					
10	0.8					
10	0.9					

(a) Use a spreadsheet to compute the variance for the binomial distributions in the above spreadsheet.
(b) State the value of p that results in the largest variance.
(c) State the value(s) of p that result in the smallest variance.

7. *Generating binomially distributed random variables.* Assume a production process that, historically, produces 20% defectives. We will select a sample of 8 items from the output of this production process. Let X denote the number of defectives obtained in the sample of 8 items. Assuming independent trials, X is a binomially distributed random variable with $n = 8$ and $p = 0.20$. This binomial distribution is obtained using Excel and is shown in Spreadsheet 7-5.

If we repeat this sampling process any number of times, say 20 times, and list the number of defectives, X, obtained in each sample of size 8, then we have generated values of, or equivalently, *simulated* a binomially distributed random variable, X. Spreadsheet 7-5 gives the 20 values of X in the column labeled X. Each of these 20 values of X (20 samples) represents the number of defectives obtained from each sample of size $n = 8$.

SPREADSHEET 7-5

X		Simulated results				Binomial distribution	
		x	f	%f		X	P(X)
2		0	5	0.25		0	0.167772
5		1	7	0.35		1	0.335544
0		2	3	0.15		2	0.293601
1		3	4	0.2		3	0.146801
0		4	0	0		4	0.045875
3		5	1	0.05		5	0.009175
3		6	0	0		6	0.001147
3		7	0	0		7	8.19E-05
1		8	0	0		8	2.56E-06
3			20	1			1
1							
2							
0							
1							
0							
0							
1							
1							
1							
2							

Observe in Spreadsheet 7-5 that we have tabulated the 20 values of X showing the frequency of occurrence (f) of each and the corresponding percent frequency (%f) so that we can compare the *simulated results* with the *theoretical binomial distribution* given in the upper-right-hand corner of the spreadsheet. Notice that the percent frequencies (%f) reveal a similar pattern to the binomial probabilities, $P(X)$, such that larger percents (probabilities) occur at smaller x-values for both simulated results and theoretical binomial distribution. Note that the percent frequencies (%f) under **Simulated results** in Spreadsheet 7-5 were obtained by dividing each frequency (f) by the total frequency, 20.

INSTRUCTIONS

Use the following steps to generate (simulate) a binomially distributed random variable, X, as listed in the X- column of Spreadsheet 7-5.

1. Select *Tools* from the menu bar and then select *Data Analysis*. If Data Analysis does not appear, select *Add-Ins* and then select *Analysis ToolPak* and click **OK**.

2. After selecting *Data Analysis*, click the down arrow repeatedly until *Random Number Generation* appears. Click on *Random Number Generation* until it is highlighted, and then click **OK**.

3. Enter **1** in the white box next to Number of Variables to indicate that we wish to generate values for <u>one</u> random variable.

4. In the white box labeled *Number of Random Numbers*, enter **20**, because we wish to generate 20 random numbers.

5. In the white box labeled *Distribution*, click on the arrow and select *Binomial*.

6. Enter the value for p (in this example, **.20**) in the white box labeled p *Value*. Then enter the value for n, **8**, in the white box labeled *Number of Trials*.

7. Click on *Output Range* and, in its white box, enter **a2:a21** to identify the cell range that is to contain the values of the random variable being generated. Click **OK**, and the values appear in column X.

(a) Generate 30 values of the binomially distributed ($n = 8$, $p = 0.2$) random variable X for this problem. In other words, use the above instructions to simulate (repeat) this sampling process 30 times, and list the number of defectives, X, obtained in each sample of size 8.

(b) Create a table of *Simulated results* similar to that in Spreadsheet 7-5. Create the table of theoretical binomial probabilities as given in the upper-right-hand corner of Spreadsheet 7-5. Compare the percent frequencies (%f) of the simulated results to the probabilities of the theoretical binomial distribution.

8. *Generating binomially distributed random variables.* A new drug for treating some disease has been proven to be effective 70% of the time. The drug is administered to a sample of 9 patients suffering with the disease. Let the random variable X denote the number of patients cured. Simulate this sampling process 20 times, tabulate the simulated results, obtain the theoretical binomial distribution ($n = 9$, $p = 0.7$), and compare the simulated results to the theoretical binomial distribution.

7-4 The Normal Distribution

Previous sections of this chapter have discussed probability distributions where the random variables assumed discrete values. Such probability distributions are called *discrete distributions*. Probability distributions where the random variable assumes a continuous range of values are called *continuous distributions*. This section discusses a very important continuous probability distribution—the ***normal distribution***. Spreadsheet 7-6 provides graphs of two normal distributions. They have standard deviations of 1 and 2, respectively, but both have a mean of 0.

SPREADSHEET 7-6

x	y: Stdev 1	y: Stedv 2	NORMAL DISTRIBUTIONS		
-5	0.000001	0.008764			
-4.5	0.000016	0.01587			
-4	0.000134	0.026995			
-3.5	0.000873	0.043139			
-3	0.004432	0.064759			
-2.5	0.017528	0.091325			
-2	0.053991	0.120985			
-1.5	0.129				
-1	0.241				
-0.5	0.352				
0	0.398				
0.5	0.352				
1	0.241				
1.5	0.129				
2	0.053				
2.5	0.017				
3	0.004				
3.5	0.000				
4	0.000134	0.026995			
4.5	0.000016	0.01587			
5	0.000001	0.008764			

Tables of *y*-values for both distributions are given in the spreadsheet. Observe that normally distributed random variables are symmetrically dispersed about the mean, which, in both cases, is 0. Of course, the distribution with a standard deviation of 2 is more spread out than the distribution with a standard deviation of 1.

INSTRUCTIONS

Use the following steps to find:
(a) normal-curve areas located to the left of given x-values;
(b) y-coordinates of points on the normal curve corresponding to given x-values.

  For example, we consider a normal curve with a mean of 500 and a standard deviation of 100.

1. Move the dark-bordered rectangle to any cell and click f_x, , *Paste Function*, on the toolbar. A dialog box appears. Use your mouse pointer to select *Statistical* in the FUNCTION CATEGORY column and then select *Normdist* in the FUNCTION NAME column. Click **OK**, and another dialog box appears.

2. Click in the box labeled x and enter the **x-value** such that the desired *area lies to its left*. For this example, we enter **600**.

3. Click in the box labeled MEAN and enter the **mean** of the normal distribution— in this example, **500**.

4. Click in the box labeled STANDARD_DEV and enter the **standard deviation**—in this example, **100**.

5. Click in the box labeled CUMULATIVE, enter **1,** and click **OK**, and the normal-curve *area to the left of the x-value* entered in Step 2 appears in the cell with the dark-bordered rectangle. Entering **0** here returns the y-coordinate on the normal curve corresponding to the x-value entered in Step 2. This is how we obtained the y-values in Spreadsheet 7-6.

INSTRUCTIONS

Use the following steps to find x-values when the normal curve areas lying to their left are known.

  For example, we consider determining the 90th percentile for a normal curve with a mean of 500 and a standard deviation of 100. In other words, we seek the x-value when the normal-curve area located to its left is 0.9000.

1. Move the dark-bordered rectangle to any cell and click f_x, , *Paste Function*, on the toolbar. A dialog box appears. Use your mouse pointer to select *Statistical* in the FUNCTION CATEGORY column and then select *Norminv* in the FUNCTION NAME column. Click **OK**, and another dialog box appears.

2. Click in the box labeled PROBABILITY and enter **0.90**.

3. Click in the box labeled MEAN and enter **500**.

4. Click in the box labeled STANDARD_DEV, enter **100**, and click **OK**, and the 90th percentile appears in the cell with the dark-bordered rectangle. Remember, this is the *x*-value with 0.90 of the normal-curve area located to its left.

INSTRUCTIONS

Use the following steps to find the areas of the ***standard normal curve*** located to the left of given *z*-values.

  Although not needed here, we note that the standard normal curve has a mean of 0 and a standard deviation of 1. Only the *z*-values are needed to obtain normal curve areas located to their left.

  For example, we consider finding the normal curve area located to the left of $z = 1.5$.

1. Move the dark-bordered rectangle to any cell and click f_x, , ***Paste Function***, on the toolbar. A dialog box appears. Use your mouse pointer to select ***Statistical*** in the FUNCTION CATEGORY column and then select ***Normsdist*** in the FUNCTION NAME column. Click **OK**, and another dialog box appears. Note that the letter *s* in the function name denotes ***standard*** normal distribution.

2. Click in the box labeled Z, enter **1.5**, click **OK**, and the area 0.9332 appears in the cell with the dark-bordered rectangle.

INSTRUCTIONS

Use the following steps to find *z*-values when the normal-curve areas lying to their left are known.

  For example, we consider determining the 90th percentile for a standard normal curve. In other words, we seek the *z*-value when the normal-curve area located to its left is 0.9000.

1. Move the dark-bordered rectangle to any cell and click f_x, , ***Paste Function***, on the toolbar. A dialog box appears. Use your mouse pointer to select ***Statistical*** in the FUNCTION CATEGORY column and then select ***Normsinv*** in the FUNCTION NAME column. Click **OK**, and another dialog box appears. Again, note that the letter *s* in the function name denotes ***standard*** normal distribution.

2. Click in the box labeled PROBABILITY, enter **0.90**, and click **OK**, and the 90th percentile appears in the cell with the dark-bordered rectangle. Remember, this is the *z*-value with 0.90 of the normal-curve area located to its left.

140

EXERCISES

1. *Areas to the left of given x-values.* A normal distribution has a mean of 450 and a standard deviation of 50. Use a spreadsheet to find the normal-curve area located to the left of:

(a) $x = 300$ (b) $x = 325$ (c) $x = 350$ (d) $x = 385$ (e) $x = 450$ (f) $x = 470$ (g) $x = 495$

2. *Areas to the right of given x-values.* Find the normal-curve area located to the *right* of each x-value given in Exercise 1.

3. *Areas to the right of the mean.* A normal distribution has a mean of 300 and a standard deviation of 20. Use a spreadsheet to find the normal-curve area located between 300 and:

(a) $x = 310$ (b) $x = 325$ (c) $x = 340$ (d) $x = 345$ (e) $x = 350$ (f) $x = 360$

Explain why, in each case, 0.50 must be subtracted from the area derived from *Normdist*.

4. *Areas to the left of the mean.* A normal distribution has a mean of 300 and a standard deviation of 20. Use a spreadsheet to find the normal-curve area located between the x-value given below and 300:

(a) $x = 245$ (b) $x = 265$ (c) $x = 270$ (d) $x = 285$ (e) $x = 290$ (f) $x = 295$

Explain why, in each case, the area derived from *Normdist* must be subtracted from 0.50.

5. *Areas between two x-values.* A normal distribution has a mean of 300 and a standard deviation of 20. Use a spreadsheet to find the normal-curve area located between the x-values:

(a) 245 and 255   (b) 260 and 275   (c) 285 and 310   (d) 315 and 340   (e) 345 and 355

Explain why, in each case, the areas derived from *Normdist* must be subtracted.

6. *Percentiles.* A normal distribution has a mean of 400 and a standard deviation of 80. Use a spreadsheet to find the:

(a) 20th percentile (b) 30th percentile (c) 43rd percentile (d) 50th percentile
(e) 60th percentile (f) 77th percentile (g) 80th percentile (h) 95th percentile

7. *Standard normal distribution.* Use a spreadsheet to find the area to the left of each indicated z-value.

(a) $z = -3$ (b) $z = -2.24$ (c) $z = -1.78$ (d) $z = 0$ (e) $z = 1.5$ (f) $z = 2.3$ (g) $z = 3$

8. *Standard normal distribution.* Use a spreadsheet to find the z-value corresponding to the:

(a) 30th percentile (b) 45th percentile (c) 50th percentile (d) 93rd percentile

9. *Daily demand.* The daily demand for pancake batter at Patty's Pancake House is normally distributed, with a mean daily demand of 600 pounds and a standard deviation of 80 pounds. Use a spreadsheet to answer each of the following.

(a) Find the probability that daily demand is less than 700 pounds.

(b) Find the probability that daily demand is greater than 710 pounds.

(c) Find the probability that daily demand is between 715 and 830 pounds.

(d) How many pounds of batter should be kept on hand in order to satisfy daily demand 70% of the time?

(e) How many pounds of batter should be kept on hand in order to satisfy daily demand 90% of the time?

(f) How many pounds of batter should be kept on hand in order to satisfy daily demand 99% of the time?

10. *Call options.* The following spreadsheet reveals an investor's profit per share (in dollars), y, on a call option as a function of the stock's price per share (in dollars), x.

x	y
0	-20
5	-20
10	-20
15	-20
20	-20
25	-20
30	0
35	20
40	40
45	60
50	80

(a) On what interval of x-values does the investor lose $20 per share?

(b) On what interval of x-values does the investor lose money?

(c) On what interval of x-values does the investor make a profit?

(d) If the stock price is normally distributed with a mean of $25 and a standard deviation of $5, use a spreadsheet to determine the probability that the investor loses $20 per share.

(e) If the stock price is normally distributed with a mean of $25 and a standard deviation of $5, use a spreadsheet to determine the probability that the investor loses money.

(f) If the stock price is normally distributed with a mean of $25 and a standard deviation of $5, use a spreadsheet to determine the probability that the investor makes a profit.

(g) If the stock price is normally distributed with a mean of $25 and a standard deviation of $5, use a spreadsheet to determine the probability that the investor makes a profit of at least $20 per share.

11. *Call options.* Consider the call option of Exercise 10. Assume the stock price is normally distributed with a mean of $20 per share and a standard deviation of $5.

(a) Use a spreadsheet to determine the probability that the investor loses $20 per share.

(b) Use a spreadsheet to determine the probability that the investor loses money.

(c) Use a spreadsheet to determine the probability that the investor makes a profit.

(d) Use a spreadsheet to determine the probability that the investor makes a profit of at least $20 per share.

12. *Plotting normal curves.* Use the x-values -10, -9, -8, -7, . . . , 10 to create a table of x-
and y-values for two normal curves that have standard deviations of 2 and 4, respectively,
but that both have means of 0.
  (a) Create the corresponding graphs.
  (b) Identify the graph with the greater standard deviation.

13. *Generating normally distributed random variables.* Here we show how to generate 20
numbers randomly selected from a normal distribution with a mean 40 and a standard
deviation of 3. Additionally, we provide a second set of instructions showing how to
obtain a histogram for a column of data values. As an example, we use the column of
normally distributed numbers generated in Spreadsheet 7-7. The histogram is also
included in Spreadsheet 7-7.

SPREADSHEET 7-7

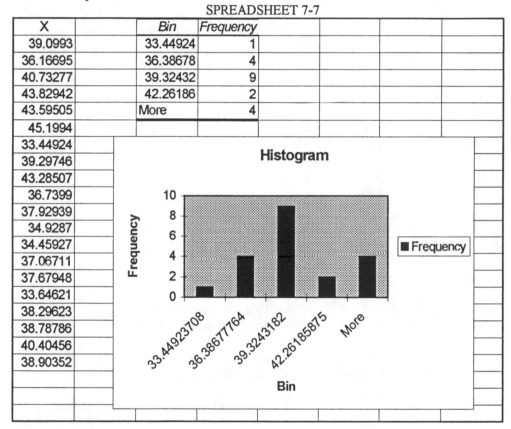

X		Bin	Frequency			
39.0993		33.44924	1			
36.16695		36.38678	4			
40.73277		39.32432	9			
43.82942		42.26186	2			
43.59505		More	4			
45.1994						
33.44924						
39.29746						
43.28507						
36.7399						
37.92939						
34.9287						
34.45927						
37.06711						
37.67948						
33.64621						
38.29623						
38.78786						
40.40456						
38.90352						

INSTRUCTIONS

Use the following steps to generate a set of numbers randomly selected from a normal
distribution with a given mean and standard deviation. In this example, we generate
20 numbers from a normal distribution with a mean of 40 and a standard deviation
of 3.

1. Select *Tools* from the menu bar and then select *Data Analysis*. If Data Analysis does not appear, select *Add-Ins*. Then select *Analysis ToolPak* and click *OK*.

2. After selecting *Data Analysis*, click the down arrow repeatedly until *Random Number Generation* appears. Click on *Random Number Generation* until it is highlighted, and then click **OK**.

3. Enter **1** in the white box next to Number of Variables to indicate that we wish to generate values for <u>one</u> random variable.

4. In the white box labeled *Number of Random Numbers*, enter **20**, because we wish to generate 20 random numbers.

5. In the white box labeled *Distribution*, click on the arrow and select *Normal*.

6. In the white box next to *Mean*, enter the mean of the normal distribution, **40**. Then enter the standard deviation, **3**, in the white box next to *Standard deviation*.

7. Click on *Output Range* and, in its white box, enter **A2:A21** to identify the cell range that is to contain the values of the random variable being generated. Click **OK**, and the values appear in column X.

INSTRUCTIONS

Use the following steps to **OBTAIN A HISTOGRAM** of a column of data values. We will use the following instructions to obtain a histogram of the normally distributed values of X generated in Spreadsheet 7-7. The resulting histogram also appears in Spreadsheet 7-7.

1. Select *Tools* from the menu bar and then select *Data Analysis*. If Data Analysis does not appear, select *Add-Ins*. Then select *Analysis ToolPak* and click **OK**.

2. After selecting *Data Analysis*, click the down arrow repeatedly until *Histogram* appears. Click on *Histogram* until it is highlighted, and then click **OK**.

3. In the white box next to *Input Range*, enter **A1:A21**, the cell range that contains the data. Note that our cell range includes the column label X, so we click the small white square next to *Label* to indicate that our entered cell range includes a label.

4. In the box next to *Output Range*, enter the cell that marks the upper-left-hand corner of the range where we want the output to appear on our spreadsheet. In this case, we enter **C1**.

5. Click on the small white box next to *Chart Output* (located at the bottom of the dialog box) *to obtain a histogram*. Click **OK**. Note that cell C1 marks the beginning of our output that includes a frequency distribution of the data values with a histogram appearing to its right. Increase the size of the histogram and note that, for our example, its shape should resemble (perhaps imperfectly) the normal curve. If desired, change the location of the histogram.

(a) Generate 30 numbers from a normal distribution with a mean of 50 and a standard deviation of 4. Note that this is the same as selecting a random sample of 30 numbers from a normal distribution having a mean of 50 and a standard deviation of 4. Create a histogram of the generated numbers and comment on its resemblance to the normal curve.

(b) Generate 35 numbers from a normal distribution with a mean of 60 and a standard deviation of 2.5. Note that this is the same as selecting a random sample of 35 numbers from a normal distribution having a mean of 60 and a standard deviation of 2.5. Create a histogram of the generated numbers and comment on its resemblance to the normal curve.

14. *Simulating product lifetimes.* A certain brand of light bulb lasts, on the average, 10,000 hours, with a standard deviation of 500 hours. The bulb lifetimes are normally distributed. Simulate the lifetimes of 50 such bulbs and obtain a histogram of the results. Note that this is the same as generating 50 numbers from a normal distribution having a mean of 10,000 and a standard deviation of 500.

15. *Simulating product lifetimes.* The length of life of a certain auto battery is normally distributed with a mean lifetime of 3 years and a standard deviation of 0.5 year. Simulate the lifetimes of 40 such batteries and obtain a histogram of the results.

16. *Simulating product fill.* A dairy distributes most of its milk in half-gallon (64-ounce) containers. The machine filling these containers is imperfect (as are most things). It distributes, on the average, 64 ounces of milk per half-gallon container, with a standard deviation of 0.5 ounce. If the amount poured into each half-gallon container is a normally distributed random variable, simulate the filling of 45 such containers and obtain a histogram of the results.

17. *Simulating exam grades.* Final exam grades on a particular standardized exam are normally distributed with a mean grade of 60 and a standard deviation of 10. Simulate the grades obtained on this examination by a random sample of 50 students and obtain a histogram of the results.

18. *Generating a bimodal distribution.* Generate 20 numbers (in cells **A2:A21**) from a normal distribution with a mean of 5 and a standard deviation of 0.5. Then generate 20 numbers (in cells **A22:A41**) from a normal distribution with a mean of 9 and a standard deviation of 0.5. Obtain a histogram of the combined set of 40 numbers (enter the *Input Range* **A1:A41** and click *Label* to include the label *x*) and note that it has two peaks. Such a distribution is called a *bimodal distribution*. Spreadsheet 7-8 provides an example of these results. A distribution with two peaks usually indicates that the data come from two different sources. In this case, one source was a normal distribution with a mean of 5

x		Bin	Frequency			
4.849884		3.908206	1			
4.361158		5.042444	13			
5.122129		6.176682	6			
5.638237		7.31092	0			
5.599175		8.445158	4			
5.866567		9.579396	13			
3.908206		More	3			
4.882909						
5.547511						
4.45665						
4.654898						
4.154784						
4.076545						
4.511185						
4.613246						
3.941034						
4.716038						
4.797976						
5.067427						
4.817254						
9.211188						
8.911238						
9.044639						
8.955592						
8.224526						
8.663295						
9.560801						
9.10936						
9.419036						
8.751782						
8.364946						
9.795847						
9.236988						
8.886288						
9.253553						
8.234681						
9.724053						
8.049546						
8.787556						
10.71363						

Histogram

and a standard deviation of 0.5; the other source was normal distribution with a mean of 8 and a standard deviation of 0.5. The means differ by an amount large enough so that there are two distinct peaks to the distribution of the combined data.

19. *Generating a bimodal distribution.* Generate 20 numbers (in cells **A2:A21**) from a normal distribution with a mean of 4 and a standard deviation of 0.3. Then generate 20 numbers (in cells **A22:A41**) from a normal distribution with a mean of 5.8 and a standard deviation of 0.3. Obtain a histogram of the combined set of 40 numbers. Comment on the histogram's shape and explain the result.

20. *Generating a bimodal distribution.* Generate 30 numbers (in cells **A2:A31**) from a normal distribution with a mean of 7 and a standard deviation of 0.4. Then generate 30 numbers (in cells **A22:A61**) from a normal distribution with a mean of 9 and a standard deviation of 0.4. Obtain a histogram of the combined set of 60 numbers. Comment on the histogram's shape and explain the result.

21. *Simulating the mixing of output from two production processes.* The diameters of diskettes coming from production process 1 are normally distributed with a mean of 3.50 inches and a standard deviation of 0.005 inch; the diameters of diskettes coming from production process 2 are normally distributed with a mean of 3.52 inches and a standard deviation of 0.005 inch. Simulate the mixing of a sample of 30 diskettes from production process 1 with a sample of 30 diskettes from production process 2. Comment on the shape of the histogram.

22. *Simulating the mixing of two sets of examination scores.* Examination scores of students from school 1 are normally distributed with a mean of 70 and a standard deviation of 2. Examination scores of students from school 2 are normally distributed with a mean of 80 and a standard deviation of 2. Simulate the mixing of a random sample of 30 students from each school. Comment on the shape of the histogram.

23. *Overcontrol of a production process.* A production process is set to produce metal shafts having a mean length of 3.000 inches with a tolerance of ± 0.015 inch. This means that a shaft is considered acceptable as long as its length lies within the interval $3.000 inches \pm 0.015 inch$. Equivalently, if x denotes the shaft length, then a shaft is acceptable as long as $2.985'' \leq x \leq 3.015''$. Overcontrol of a production process occurs when a quality control supervisor notices that shaft lengths are at, say 2.99 inches, which is within the acceptable range. However, in an attempt at perfection, the supervisor resets the production process to produce shaft lengths of, say 3.01 inches. The result is a probability distribution of shaft lengths that has greater variability than the one with the original setting. We simulate this situation by assuming that the shaft lengths are normally distributed with a standard deviation of 0.005 inch. However, our simulation includes the mixing of two sets of data, one with a mean of 2.99 inches and the other with a mean of 3.01 inches. Our simulation includes 20 data values for each set of data and is given in Spreadsheet 7-9

Observing the histogram in Spreadsheet 7-9, note that the distribution has a tendency to be flat with minimal (if any) peak in the vicinity of the center. This is due to the fact that the means of the two sets of data do not differ by an amount large enough to create

x		Bin	Frequency			
2.988499		2.979082	1			
2.983612		2.987091	8			
2.991221		2.9951	7			
2.996382		3.003109	7			
2.995992		3.011118	9			
2.998666		3.019127	7			
2.979082		More	1			
2.988829						
2.995475						
2.984566						
2.986549						
2.981548						
2.980765						
2.985112						
2.986132						
2.97941						
2.98716						
2.98798						
2.990674						
2.988173						
3.012112						
3.009112						
3.010446						
3.009556						
3.002245						
3.006633						
3.015608						
3.011094						
3.01419						
3.007518						
3.003649						
3.017958						
3.01237						
3.008863						
3.012536		StDev 1	0.005686			
3.002347		StDev2	0.006266			
3.017241		StDevCo	0.012904			
3.000495						
3.007876						
3.027136						

Histogram

two separate peaks. If they did, the result would have been a bimodal distribution. Instead, the *combined distribution is more spread out so that its variability is greater than that of the original*. Evidence of this is given in Spreadsheet 7-9, where below the histogram are three standard deviations: *StDev1* is the standard deviation of the first set of data, *StDev2* is the standard deviation of the second set of data, and *StDevCom* is the standard deviation of the combined set of data. Note that the standard deviation of the combined set of data is more than double that of either set. This illustrates that *overcontrol of a production process results in increased variability* of the final output.

INSTRUCTIONS

Determining the Standard Deviation
Finally, we note that we determined the standard deviation (StDev1) of the first set of data by moving the dark-bordered rectangle to the cell that is to contain the standard deviation and entering the Excel formula =STDEV(A2:A21). Equivalently, we could click f_x, *Paste Function*, on the toolbar, select *Statistical*, and then select *StDev* and enter A2:A21, the range of cells containing the first set of data values. Of course, we enter the cell range **A22:A41** to determine the standard deviation (StDev2) of the second set of data; we enter the cell range **A2:A41** to determine the standard deviation (StDevCom) of the combined set of data.

(a) Repeat this simulation on your spreadsheet. Although your results will probably differ from those of Spreadsheet 7-9, the overall shape of the histogram and the conclusion regarding variability should be similar.
(b) Repeat this simulation using 30 data values for each set of data.

24. *Overcontrol of a production process.* Simulate a situation similar to Exercise 23 by assuming that the shaft lengths are normally distributed with a standard deviation of 0.004 inch. Let your simulation include the mixing of two sets of data, one with a mean of 2.98 inches and the other with a mean of 3.00 inches. Use 25 data values for each set of data. Comment on the shape of the resulting histogram and the variability of the combined data.

25. *Overcontrol of a production process.* Repeat Exercise 24 using 30 data values for each set of data.

26. *Overcontrol of a production process.* Repeat Exercise 24 using 35 data values for each set of data.

CHAPTER EIGHT

Derivatives

8-1 Limits

When we apply the concept of limit, we examine what happens to the y-values of a function $f(x)$ as x gets closer and closer to (but does not reach) some particular number, called a. If the y-values also get closer and closer to a *single number*, L, then the number L is said to be the limit of the function as x approaches a. Thus, we say that L **is the limit of** $f(x)$ **as** x **approaches** a. This is written in mathematical shorthand as

$$\lim_{x \to a} f(x) = L$$

where the symbol $\to$ stands for the word "approaches." If the y-values of the function do <u>not</u> get closer and closer to a single number as x gets closer and closer to a, then the function has <u>no limit</u> as x approaches a.

Spreadsheet 8-1 gives the y-values for the function

$$f(x) = \frac{2(x^2 - 9)}{x - 3}$$

as x approaches 3. Specifically, Table A gives the y-values as x approaches 3 from the *left* and Table B gives the y-values as x approaches 3 from the *right*. Note that in Table

SPREADSHEET 8-1

TABLE A				TABLE B	
x	2(x^2 - 9)/(x - 3)			x	2(x^2 - 9)/(x - 3)
2.9	11.8			3.1	12.2
2.99	11.98			3.01	12.02
2.999	11.998			3.001	12.002
2.9999	11.9998			3.0001	12.0002
2.99999	11.99998			3.00001	12.00002

A, the y-values approach 12 as the x-values approach 3 from the left and, that in Table B, the y-values also approach 12 as the x-values approach 3 from the right. In other words, as we move down the columns of Table A, the *y-values differ from each other by lesser and lesser amounts* so that they approach a single value, 12. The same occurs as we move down Table B. The fact that the *y*-values from both the left and right of $x = 3$ approach a *single value*, 12, indicates that

$$\lim_{x \to 3} \frac{2(x^2 - 9)}{x - 3} = 12 .$$

No Limit

To provide an example of a function that *does not have a limit* at a point, Spreadsheet 8-2 gives the *y*-values for the function

$$f(x) = \frac{1}{x^2}$$

as *x* approaches 0. Studying either Table A or B, note that the *y*-values do *not* approach a

SPREADSHEET 8-2

TABLE A				TABLE B		
X	1/X^2			x	1/x^2	
-0.1	100			0.1	100	
-0.01	10000			0.01	10000	
-0.001	1000000			0.001	1000000	
-0.0001	100000000			0.0001	100000000	
-0.00001	10000000000			0.00001	10000000000	

single value as in Spreadsheet 8-1 but, rather, increase without bound as the *x*-values approach 0. In other words, as we move down Table A or B, the *y*-values differ from each other by greater and greater amounts. Thus, this function has *no limit* as *x* approaches 0. Other examples where functions do not have limits are provided in the exercises.

INSTRUCTIONS

Use the following instructions to create tables and graphs similar to those in Spreadsheet 8-1.

1. Type Labels

1.1 As an example, we will create a table of *x*- and *y*-values for the function featured in Spreadsheet 8-1. Thus, we will use Column A for the *x*-values and Column B for the *y*-values of the function $f(x) = \frac{2(x^2 - 9)}{x - 3}$.

1.2 Once you have a blank worksheet, use the mouse to move your pointer to cell A1 and click the left mouse button to make that cell the active cell. A rectangle with a dark border should appear around cell A1. Use the spacebar to move past the middle of the cell and type **TABLE A** as illustrated in Spreadsheet 8-1. Use either the mouse or the arrow keys to move to cell A3.

1.3 Use the spacebar to move to the middle of the cell and type **x**. This labels the remaining portion of Column A. Now, we will label the corresponding section of Columns B with the equation $y = 2(x^2 - 9) / (x - 3)$. Use either the mouse or the arrows on the keyboard to move the dark-bordered rectangle around cell B3. Type **2(x^2 - 9)/(x - 3)** in cell B3.

Note: *Widening a Column*
To *widen Column B* so that the formula fits in a single column, *move the mouse pointer to the right border of the Column B heading at the top of the spreadsheet so that the pointer becomes a black two-way arrow and double-click the right border of Column B.* The column will widen to accommodate its contents. This procedure is also used to widen a column to accommodate a number with many digits. For example, **#####** appearing in a cell means that the column is too narrow to accommodate the desired number. Using the double-click procedure will widen the column so that the number will display.

2. Create a Table of *x*- and *y*-values
2.1 Beginning with cell A4, type the *x*-values **2.9** through **2.99999** into the section of Column A as illustrated in Spreadsheet 8-1.

2.2 Now, we *enter a formula* to compute the corresponding *y*-values for the equation $y = (x^2 - 9) / (x - 3)$ in Column B. Move the dark-bordered rectangle to cell B4, type the formula **=2*(A4^2-9)/(A4-3)**, and press **Enter**. Note that the symbol * means multiplication and the symbol ^ means exponentiation. The formula **= 2*(A4^2-9)/(A4-3)** entered in cell B4 returns to cell B4 the *y*-value ($y = 11.8$) corresponding to the *x*-value ($x = 2.9$) of cell A4. *Note that a formula must always begin with an equals (=) sign.*

2.3 Now, we *copy the formula* down through cell B8. Move the dark-bordered rectangle to cell B4. Use the mouse to move the pointer to the small black box (called a handle) at the lower right corner of cell B4. The mouse pointer becomes a thick black plus. Click the mouse button without releasing it and drag the mouse pointer down to cell B8. Release the mouse button at cell B8, and cells B4 to B8 will contain the formula values— in other words, the *y*-values corresponding to the *x*-values of Column A.

2.4 Repeat this procedure for Table B.

EXERCISES

1. Use Excel to create and complete the tables shown in the following spreadsheet. Use the results to estimate the indicated limit (if it exists). $\lim_{x \to 2} \dfrac{3(x^2 - 4)}{x - 2}$

x	3(x^2-4)/(x-2)		x	3(x^2-4)/(x-2)	
1.9			2.1		
1.99			2.01		
1.999			2.001		
1.9999			2.0001		
1.99999			2.00001		

2. Use Excel to create and complete the tables shown in the spreadsheet below. Use the results to estimate the following limit (if it exists). $\lim\limits_{h \to 0} \dfrac{\sqrt{1+h}-1}{h}$

h	((1+h)^.5 - 1)/h		h	((1+h)^.5 - 1)/h	
-0.5			0.5		
-0.4			0.4		
-0.3			0.3		
-0.2			0.2		
-0.1			0.1		
-0.01			0.01		
-0.001			0.001		
-0.0001			0.0001		
-0.00001			0.00001		
-0.000001			0.000001		

The formula $\dfrac{\sqrt{1+h}-1}{h}$ can be expressed as indicated at the top of the spreadsheet. In other words, assuming that the formula will be entered in cell B2, we could enter either =((1+A2)^.5-1)/A2 or =((1+A2)^(1/2)-1)/A2. However, Excel's *square root function* can be used to express the formula as =(SQRT(1+A2)-1)/A2. A useful exercise is to try all three ways and compare the results. They should be the same.

3. Use Excel to create and complete the tables shown in the spreadsheet below. Use the results to explain why the following limit does *not* exist. $\lim\limits_{x \to 0} \dfrac{|x|}{x}$

x	|x|/x		x	|x|/x	
-0.2			0.2		
-0.1			0.1		
-0.01			0.01		
-0.001			0.001		
-0.0001			0.0001		

Assuming that the formula will be entered in cell B2, Excel's *absolute value function* can be used to express the formula as =ABS(A2)/A2.

4. Use Excel to create and complete the tables shown in the spreadsheet below. Use the results to explain why the following limit does *not* exist. $\lim\limits_{x \to 0} \dfrac{1}{x}$

x	1/x		x	1/x		
-0.1			0.1			
-0.01			0.01			
-0.001			0.001			
-0.0001			0.0001			
-0.00001			0.00001			
-0.000001			0.000001			

5. The number *e* is defined as either of the following limits:

$$e = \lim_{x \to \infty} \left(1 + \frac{1}{x}\right)^x \qquad \text{or} \qquad e = \lim_{x \to 0} (1 + x)^{1/x}$$

Use Excel to create and complete the tables shown in the spreadsheet below to verify that both formulas give the same approximation of *e*, which, to fifteen decimal places, is

$$e = 2.7 \ 1828 \ 1828 \ 45 \ 90 \ 45$$

where the spaces are placed so that the digits are easy to remember.

x	(1+1/x)^x		x	(1+x)^(1/x)		
1			1			
10			0.1			
100			0.01			
1000			0.001			
10000			0.0001			
100000			0.00001			
1000000			0.000001			
10000000			0.0000001			
100000000			0.00000001			

8-2 Instantaneous Rate of Change: Numerical Approximation

We have learned that the average rate of change of a function is the rate of change between two points, whereas the instantaneous rate of change is the rate of change at a single point. We now show how to approximate the instantaneous rate of change by numerical computation.

To approximate numerically the instantaneous rate of change of a function at a point x, we compute the average rate of change between x and $x + h$ for successively smaller values of h. If the average rates of change approach a single value as h gets smaller and smaller, then that single value is the instantaneous rate of change.

Graphical Interpretation
Graphically, each average rate of change is the slope of the secant line between points x and $x + h$ on the graph of $f(x)$.

As an illustrative example, we consider

$$f(x) = 10x^2$$

where $f(x)$ denotes the distance (in miles) traveled by a car after x hours. We use Spreadsheet 8-3 below to numerically approximate the instantaneous rate of change of $f(x)$ at $x = 3$.

SPREADSHEET 8-3

x	h	x + h	f(x + h)	f(x)	[f(x+h)-f(x)]/h
3	0.1	3.1	96.1	90	61
3	0.01	3.01	90.601	90	60.1
3	0.001	3.001	90.06001	90	60.01
3	0.0001	3.0001	90.006	90	60.001
3	0.00001	3.00001	90.0006	90	60.0001

The rightmost column of Spreadsheet 8-3, labeled **[f(x + h) - f(x)]/h**, gives the average rates of change of $f(x)$ between the points $x = 3$ and $x = 3 + h$ for successively smaller values of h. Graphically, these are the slopes of the secant lines between $x = 3$ and $x = 3 + h$ on the graph of $f(x)$. Note that as h gets smaller and smaller, the average rates of change are approaching 60. At this point, it appears that 60 is the instantaneous rate of change at $x = 3$. However, because we chose *positive values for h*, the average rates of change give the slopes of secant lines to the *right of x = 3*.

We must check that the slopes of the secant lines to the *left of x = 3* also approach 60. Spreadsheet 8-4 gives the average rates of change for such secant lines by using *negative values for h*.

155

x	h	x + h	f(x + h)	f(x)	[f(x+h)-f(x)]/h	
3	-0.1	2.9	84.1	90	59	
3	-0.01	2.99	89.401	90	59.9	
3	-0.001	2.999	89.94001	90	59.99	
3	-0.0001	2.9999	89.994	90	59.999	
3	-0.00001	2.99999	89.9994	90	59.9999	

Studying the average rates of change given in the rightmost column of Spreadsheet 8-4, note that the average rates of change (slopes of the secant lines to the left of $x = 3$) also are approaching 60. Because the slopes of the secant lines *both to the right and to the left of* $x = 3$ approach a single value of 60, this indicates that 60 is the instantaneous rate of change of $f(x)$ at $x = 3$. In other words, $\lim_{h \to 0} \dfrac{f(x+h) - f(x)}{h} = 60$.

INSTRUCTIONS

Use the following instructions to create tables of difference quotients similar to those in Spreadsheet 8-3.

1. Type Labels
1.1 As an example, we will create a table of difference quotients to approximate numerically the instantaneous rate of change of

$$f(x) = 10x^2$$

at $x = 3$.

Once you have a blank worksheet, use the mouse to move your pointer to cell A1 and click the left mouse button to make that cell the active cell. A rectangle with a dark border should appear around cell A1. Use the spacebar to move to the middle of the cell and type **x**. This labels column A. Use the following instructions to label columns B through F.

1.2 In cell B1, type **h**.

1.3 In cell C1, type **x+h**.

1.4 In cell D1, type **f(x+h)**.

1.5 In cell E1, type **f(x)**.

1.6 In cell F1, type **[f(x+h)-f(x)]/h**. Using the mouse, move the pointer to the Column F heading and *double-click the right border of column F* heading to widen the column so that **[f(x+h)-f(x)]/h** fits in a single column.

2. Create a Table of Difference Quotients

2.1 Type **3** in cells A2 through A6. Then type **0.1, 0.01, 0.001, 0.0001,** and **0.00001** in cells B2 through B6, respectively.

2.2 Now, we *enter a formula* to compute the $x + h$ values in Column C. Move to cell C2, type the formula **=A2+B2**, and press **Enter**. The value, 3.1, should appear in cell C2.

2.3 Now, we *copy the formula* down through cell C6. Move the dark-bordered rectangle to cell C2. Use the mouse to move the pointer to the small box (called a *handle*) at the lower right corner of cell C2. The mouse pointer becomes a thick black plus. Click the mouse button without releasing it and drag the mouse pointer down to cell C6. Release the mouse button at cell C6, and cells C3 to C6 will contain the formula values.

2.4 Now, we create the column of $f(x+h)$ values. Move the pointer to cell D2 and *enter the formula* **=10*C2^2**. Note that this formula squares the value in cell C2 and multiplies the result by 10. We *copy the formula* down through cell D6 by moving the dark-bordered rectangle to cell D2, using the mouse to point to the *handle* (the small black box at the lower right corner of cell D2) so that the mouse pointer becomes a thick black plus, clicking the mouse button without releasing it, and dragging the mouse pointer down to cell D6. Upon our releasing the mouse button at cell D6, the formula values appear in cells D3 through D6. Note that this means that the respective values of Column C are squared and then multiplied by 10 as indicated by the formula.

2.5 We can either type 90 in columns E2 through E6 or *enter the formula* **=10*A2^2** in column E2 and *copy the formula* down through cell E6.

2.6 Finally, we create the column of difference quotients, $[f(x+h)-f(x)]/h$, by moving to Column F2 and *entering the formula* **=(D2-E2)/B2**. Then we *copy the formula* down through cell F6.

EXERCISES

1. Approximate the instantaneous rate of change of $f(x) = 10x^2$ at $x = 4$ by creating and completing the tables given in Spreadsheets 8-5 and 8-6. Spreadsheet 8-5 considers the secant lines to the right of $x = 4$, and Spreadsheet 8-6, containing *negative h-values*, considers secant lines to the left of $x = 4$.

SPREADSHEET 8-5

x	h	x+h	f(x+h)	f(x)	[f(x+h)-f(x)]/h	
4	0.1					
4	0.01					
4	0.001					
4	0.0001					
4	0.00001					

SPREADSHEET 8-6

x	h	x+h	f(x+h)	f(x)	[f(x+h)-f(x)]/h	
4	-0.1					
4	-0.01					
4	-0.001					
4	-0.0001					
4	-0.00001					

2. Draw the graphs that show the secant lines corresponding to the difference quotients of Spreadsheets 8-5 and 8-6.

3. Approximate the instantaneous rate of change of $f(x) = 10x^2$ at $x = 5$ by creating spreadsheets similar to those of Exercise 1.

4. Approximate the instantaneous rate of change of $f(x) = 10x^2$ at $x = 6$ by creating spreadsheets similar to those of Exercise 1.

5. Approximate the instantaneous rate of change of $f(x) = \dfrac{1}{x^2}$ at $x = 2$ by creating spreadsheets similar to those of Exercise 1.

6. Approximate the instantaneous rate of change of $f(x) = x^3$ at $x = 2$ by creating spreadsheets similar to those of Exercise 1.

7. Approximate the instantaneous rate of change of $f(x) = x^3 - 4x + 3$ at $x = 2$ by creating spreadsheets similar to those of Exercise 1.

8. Approximate the instantaneous rate of change of $f(x) = \dfrac{1}{x-3}$ at $x = 5$ by creating spreadsheets similar to those of Exercise 1.

8-3 Differentiability: A Numerical Perspective

A function is **differentiable** at a point if its derivative exists at that point. To examine this concept from a numerical perspective, we present two spreadsheets illustrating tables of difference quotients for the function

$$f(x) = \sqrt{x}$$

or, equivalently,

$$f(x) = x^{1/2}.$$

Spreadsheets 8-7 and 8-8 give the difference quotients for the above function at $x = 4$. Spreadsheet 8-7, with *positive h-values*, gives the difference quotients to the *right* of $x = 4$. Studying the difference quotient column, note that as we move down the column, the difference quotients differ from each other by lesser and lesser amounts so that they approach a single value—in this case, 0.25.

SPREADSHEET 8-7

x	h	x+h	f(x+h)	f(x)	[f(x+h)-f(x)]/h
4	0.1	4.1	2.024846	2	0.24845673
4	0.01	4.01	2.002498	2	0.24984395
4	0.001	4.001	2.00025	2	0.24998438
4	0.0001	4.0001	2.000025	2	0.24999844
4	0.00001	4.00001	2.000002	2	0.24999984

SPREADSHEET 8-8

x	h	x+h	f(x+h)	f(x)	[f(x+h)-f(x)]/h
4	-0.1	3.9	1.974842	2	0.25158234
4	-0.01	3.99	1.997498	2	0.25015645
4	-0.001	3.999	1.99975	2	0.25001563
4	-0.0001	3.9999	1.999975	2	0.25000156
4	-0.00001	3.99999	1.999997	2	0.25000016

Spreadsheet 8-8, with *negative h-values*, gives the difference quotients to the *left* of $x = 4$. Studying the difference quotient column, note that as we move down the column, the difference quotients differ from each other by lesser and lesser amounts so that they approach a single value—in this case, 0.25. The fact that the difference quotients both from the right and from the left of $x = 4$ approach the single value 0.25 indicates that the derivative at $x = 4$ exists and is equal to 0.25.

x	h	x+h	f(x+h)	f(x)	[f(x+h)-f(x)]/h
0	0.1	0.1	0.316228	0	3.16227766
0	0.01	0.01	0.1	0	10
0	0.001	0.001	0.031623	0	31.6227766
0	0.0001	0.0001	0.01	0	100
0	0.00001	0.00001	0.003162	0	316.227766

Spreadsheet 8-9 gives the difference quotients for the above function at $x = 0$. Studying the difference quotient column, note that as we move down the column, the difference quotients differ from each other by greater and greater amounts so that they do *not* approach a single value. The fact that the difference quotients differ from each other by greater and greater amounts so that they do *not* approach a single value is indicates that the derivative does *not* exist at $x = 0$.

Of course, these results can be confirmed by finding the derivative and substituting in the appropriate x-value as follows:

$$f'(x) = \frac{1}{2\sqrt{x}}$$

$$f'(4) = \frac{1}{2\sqrt{4}} = \frac{1}{4}.$$

$$f'(0) = \frac{1}{2\sqrt{0}} = \frac{1}{0}, \text{ which is undefined.}$$

EXERCISES

1. Create spreadsheets to determine whether or not the function $f(x) = x^{1/2}$ is differentiable at $x = 1$. If the function is differentiable at $x = 1$, give the value of the derivative. Confirm this result by finding the derivative and substituting in $x = 1$.

Note: If *not* using Excel's *square root function*, be certain to enclose the fractional exponent inside parentheses when entering the formula in the spreadsheet. In other words, to raise the contents of cell A2 to the $\frac{1}{2}$ power, enter the formula =A2^(1/2).

2. Create spreadsheets to determine whether or not the function $f(x) = x^{1/2}$ is differentiable at $x = 9$. If the function is differentiable at $x = 9$, give the value of the derivative. Confirm this result by finding the derivative and substituting in $x = 9$.

3. Create spreadsheets to determine whether or not the function $f(x) = x^{1/3}$ is differentiable at $x = 1$. If the function is differentiable at $x = 1$, give the value of the derivative. Confirm this result by finding the derivative and substituting $x = 1$.

Note: Be certain to enclose the fractional exponent inside parentheses when entering the formula in the spreadsheet. In other words, to raise the contents of cell A2 to the $\frac{1}{3}$ power, enter the formula =A2^(1/3).

4. Create spreadsheets to determine whether or not the function $f(x) = x^{1/3}$ is differentiable at $x = 0$. If the function is differentiable at $x = 0$, give the value of the derivative. Confirm this result by finding the derivative and substituting $x = 0$.

5. Create spreadsheets to determine whether or not the function $f(x) = x^{2/3}$ is differentiable at $x = 1$. If the function is differentiable at $x = 1$, give the value of the derivative. Confirm this result by finding the derivative and substituting $x = 1$.

6. Create spreadsheets to determine whether or not the function $f(x) = x^{2/3}$ is differentiable at $x = 0$. If the function is differentiable at $x = 0$, give the value of the derivative. Confirm this result by finding the derivative and substituting $x = 0$.

7. Create spreadsheets to determine whether or not the function $f(x) = |x|$ is differentiable at $x = 2$. If the function is differentiable at $x = 2$, give the value of the derivative. Confirm this result by finding the derivative at $x = 2$.

8. Create spreadsheets to determine whether or not the function $f(x) = |x|$ is differentiable at $x = 0$. If the function is differentiable at $x = 0$, give the value of the derivative. Confirm this result by finding the derivative at $x = 0$.

9. Create spreadsheets to determine whether or not the function $f(x) = |x - 3|$ is differentiable at $x = 3$. If the function is differentiable at $x = 3$, give the value of the derivative. Confirm this result by finding the derivative at $x = 3$.

CHAPTER NINE

Exponential and Logarithmic Functions

9-1 Exponential Growth and Decay

Exponential functions are defined by equations of the following forms:

$$\text{Exponential growth} \qquad\qquad \text{Exponential decay}$$
$$y = ab^x \qquad\qquad\qquad\qquad y = ab^{-x}$$

where a and b are constants with $a \neq 0$ and $b > 1$.   As indicated above, the *negative* sign before the exponent, x, denotes *exponential decay*, whereas *no sign* before the exponent, x, denotes *exponential growth*.

Spreadsheet 9-1 gives a table of x- and y-values of $y = 3 \cdot 2^x$ for nonnegative x-values.

SPREADSHEET 9-1

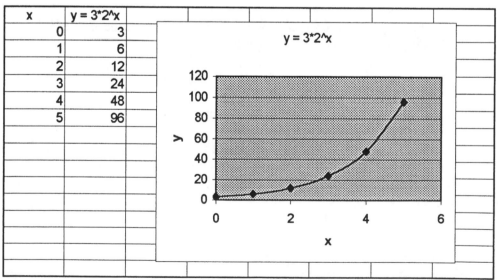

x	y = 3*2^x
0	3
1	6
2	12
3	24
4	48
5	96

The graph in Spreadsheet 9-1 typifies those of exponential growth, $y = ab^x$. Observing the tables of x- and y-values and the graph, note that the y-intercept is given by the *multiplier, a,* which in this example is 3. Scanning down the column of y-values in the spreadsheet, note that each successive y-value is the same multiple of its preceding y-value and that the multiple is b; in this case, $b = 2$. This is the distinctive property of exponential growth. Also, we note that b is called the *base* of an exponential function.

Spreadsheet 9-2 gives a table of x- and y-values of $y = 3 \cdot 2^{-x}$ for nonnegative x-values.

SPREADSHEET 9-2

x	y=3*2^-x
0	3
1	1.5
2	0.75
3	0.375
4	0.1875
5	0.09375

The graph in Spreadsheet 9-2 typifies those of exponential decay, $y = ab^{-x}$. Observing the tables of x- and y-values and the graph, note that the y-intercept is given by the *multiplier, a,* which in this example is 3. Scanning down the column of y-values in the spreadsheet, note that each successive y-value is the same multiple of its predecessor and that the multiple is b^{-1}; in this case, $b^{-1} = 2^{-1} = \dfrac{1}{2}$. This is a distinctive property of exponential decay.

Negative and Nonnegative x-values

Up to this point, we have illustrated the graphs of exponential functions for *nonnegative x-values.* Spreadsheet 9-3 gives a table of x- and y-values for $y = 3 \cdot 2^x$ for both *negative and nonnegative x-values.* Studying both table and graph of Spreadsheet 9-3, note that as the x-values get more and more negative, the y-values get smaller and smaller so that they approach 0. Thus, the x-axis is a horizontal asymptote for the exponential function.

SPREADSHEET 9-3

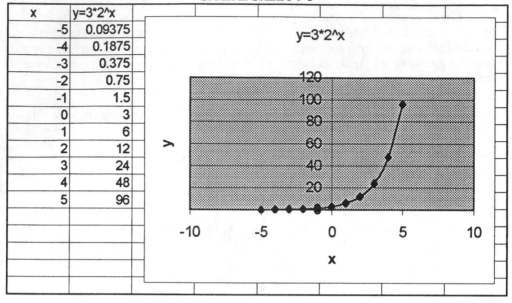

x	y=3*2^x
-5	0.09375
-4	0.1875
-3	0.375
-2	0.75
-1	1.5
0	3
1	6
2	12
3	24
4	48
5	96

$y = ae^x$ and $y = ae^{-x}$

In a homework exercise for Section 4-1, we learned that

$$e = 2.718281828$$

to nine decimal places. Because e often appears as a base for exponential functions, we note that the graph of $y = ae^x$ resembles that of exponential growth, $y = ab^x$, as typified in Spreadsheets 9-1 or 9-3. Also, the graph of $y = ae^{-x}$ resembles that of exponential decay, $y = ab^{-x}$, as typified in Spreadsheet 9-2. We have not illustrated the exponential decay graph for negative x-values. This is left for a homework exercise.

EXERCISES

1. *Exponential growth.* Create a table of x- and y-values for $y = 8 \cdot 1.7^x$. Note that this equation is of the exponential growth form $y = ab^x$ with $a = 8$ and $b = 1.7$.
   (a) Use the x-values 0, 1, 2, 3, 4, 5, 6.
      Assuming that cell A2 contains $x = 0$, type the formula **=8*1.7^A2** in cell B2.
   (b) Use Chart Wizard to create the corresponding graph.
   (c) Observe the table to determine the y-intercept and verify this result by checking the graph.

Erase only the chart by clicking inside the chart so that the handles appear on the border. If blue dashed lines are on the border, click outside the chart until the handles appear on the border. Select **Edit** from the menu bar, then select **Clear**, and then select **All**.

(d) Repeat parts (a) through (c) using the *negative x*-values -9, -8, -7, -6, -5, -4, -3, -2, -1, 0.

(e) Observe the new graph. As the *x*-values get more and more negative, the *y*-values appear to be approaching what value?

(f) Repeat parts (a) through (c) using the *x*-values -9, -8, . . ., -1, 0, 1, 2, ..., 6.

(g) Identify the horizontal asymptote.

2. *Exponential decay.* Repeat Exercise 1 for $y = 8 \cdot 1.7^{-x}$. Note that this equation is of the form $y = ab^{-x}$ with $a = 8$ and $b = 1.7$.

3. *Exponential growth.* Create a table of *x*- and *y*-values for the equations $y = e^x$ and $y = 2e^x$.

(a) Use the *x*-values -4, -3, -2, -1, 0, 1, 2, 3, 4, 5.

Note: Use the formula =**EXP(x)** for e^x and =**2*EXP(x)** for $2e^x$. In other words, if you want to get e^x corresponding to an *x*-value located in cell A2, type =**EXP(A2)** in the cell that is to contain the value of e^x.

(b) Use Chart Wizard to create the corresponding graphs.

(c) Observe the table to determine the *y*-intercepts and verify these results by checking the graph.

(d) Write the equation of the higher graph.

(e) Observe the graphs of both equations. As the *x*-values get more and more negative, the *y*-values appear to be approaching what value?

(f) Identify the horizontal asymptote.

4. *Exponential decay.* Create a table of *x*- and *y*-values for the equations $y = e^{-x}$ and $y = 2e^{-x}$.

(a) Use the *x*-values -4, -3, -2, -1, 0, 1, 2, 3, 4, 5.

Note: Use the formula =**EXP(-x)** for e^{-x} and =**2*EXP(-x)** for $2e^{-x}$.

(b) Use Chart Wizard to create the corresponding graphs.

(c) Observe the table to determine the *y*-intercepts and verify this result by checking the graph.

(d) Write the equation of the higher graph.

(e) Observe the graphs of both equations. As the *x*-values get larger and larger, the *y*-values appear to be approaching what value?

(f) Identify the horizontal asymptote.

5. **Summary:** Pencil and Paper Exercises. Based on the concepts learned in Exercises 1 through 4, draw a graph for each of the following.

(a) $y = ab^x$ with $a > 0$ and $b > 1$.   (b) $y = ae^x$ with $a > 0$.

(c) $y = ab^{-x}$ with $a > 0$ and $b > 1$.   (d) $y = ae^{-x}$ with $a > 0$.

6. ***Exponential growth.*** Create a table of x- and y-values for the equations $y = e^x$ and $y = e^{1.5x}$.

(a) Use the x-values -2, -1, 0, 1, 2, 3. Use the formula =EXP(x) for e^x and =EXP(1.5*x) for $e^{1.5x}$.

(b) Use Chart Wizard to create the corresponding graphs.

(c) Observe the table to determine the y-intercepts and verify this result by checking the graph.

(d) Write the equation of the lower graph for $x > 0$; write the equation of the lower graph for $x < 0$. At what x-value do the graphs intersect?

(e) Observe the graphs of both equations. As the x-values get more and more negative, the y-values appear to be approaching what value?

(f) Compare the two graphs. Does the multiplier, 1.5, change the general shape of the graph?

7. ***The power of exponential growth: From a penny to 10 million dollars in a month.*** We begin with a box containing a penny. Thereafter, at the end of each day, the amount of money in the box doubles so that at the end of the first day, we have 2^1 pennies; at the end of the second day, we have 2^2 pennies; at the end of the third day, we have 2^3 pennies; at the end of the xth day, we have 2^x pennies. Thus, the exponential equation $y = 2^x$ gives the formula for determining the number of pennies, y, in the box at the end of day x.

The left side of Spreadsheet 9-4 provides a format for computing the number of pennies in the box after the indicated number of days (i.e., x-values). The *right side* of Spreadsheet 9-4 provides the same format for a box that *began with 5 pennies*.

SPREADSHEET 9-4

Start with 1 penny				Start with 5 pennies	
x	y = 2^x			x	y = 5*2^x
0				0	
1				1	
7				7	
14				14	
21				21	
28				28	
30				30	

(a) Fill in the column of y-values for the *left side* of the spreadsheet. You might have to click on the right border at the top of the y-column in order to widen the column to accommodate the large numbers.

(b) How many pennies are in the box at the end of 7 days? Convert this value to dollars.

(c) How many pennies are in the box at the end of 14 days? Convert this value to dollars.

(d) How many pennies are in the box at the end of 30 days? Convert this value to dollars.

(e) Write the equation that gives the number of pennies in the box that *begins with 5 pennies*.

166

(f) Fill in the column of y-values for the **right side** of the spreadsheet. Again, you might have to click on the right border at the top of the y-column in order to widen the column to accommodate the large numbers.

(g) How many pennies are in the box at the end of 30 days? Convert this value to dollars.

8. *Writing exponential equations.* <u>Pencil and Paper Exercise</u>.   A box begins with $7. At the end of each day, the amount of money in the box doubles.

   (a) Write the equation that gives the amount of money, y, in the box after x days.

   (b) Draw a graph of the equation of part (a). Label the y-intercept with its y-coordinate.

9. *Writing exponential equations.* <u>Pencil and Paper Exercise</u>.   A box begins with $7. At the end of each day, the amount of money in the box triples.

   (a) Write the equation that gives the amount of money, y, in the box after x days.

   (b) Draw a graph of the equation of part (a). Label the y-intercept with its y-coordinate.

10. *Writing exponential equations.* <u>Pencil and Paper Exercise</u>.   A box begins with $8. At the end of each week, the amount of money in the box quadruples.

   (a) Write the equation that gives the amount of money, y, in the box after x weeks.

   (b) Draw a graph of the equation of part (a). Label the y-intercept with its y-coordinate.

11. *Writing exponential equations.* <u>Pencil and Paper Exercise</u>.   The earnings per share (EPS) of a company is currently $10. At the end of each year, the earnings per share quadruples.

   (a) Write the equation that gives the earnings per share, y, after x years.

   (b) Draw a graph of the equation of part (a). Label the y-intercept with its y-coordinate.

12. *Writing exponential equations.* <u>Pencil and Paper Exercise</u>.   The population of a small community begins with 100 people and triples at the end of each year.

   (a) Write the equation that gives the population, y, after x years.

   (b) Draw a graph of the equation of part (a). Label the y-intercept with its y-coordinate

13. *Writing exponential equations.* <u>Pencil and Paper Exercise</u>.   Eight thousand dollars is invested in a mutual fund that loses two-thirds of its value each year.

   (a) Write the equation that gives the value, y, of the investment after x years.
      Hint: If the investment loses two-thirds of its value each year, then it retains one-third. Thus, after the first year, its value is $8000(1/3)$; after the second year, its value is $8000(1/3)(1/3)$; etc.

   (b) Draw a graph of the equation of part (a). Label the y-intercept with its y-coordinate.

14. *Writing exponential equations.* <u>Pencil and Paper Exercise</u>.   Five thousand dollars is invested in a mutual fund that loses four-fifths of its value each year.

   (a) Write the equation that gives the value, y, of the investment after x years.

   (b) Draw a graph of the equation of part (a). Label the y-intercept with its y-coordinate.

15. *Exponential growth: Compound interest.* If an investment of $50 earns interest at 10% compounded annually, this means that after the first year, the $50 is increased by 10% of itself to become $50 + 50(0.10) = 50(1 + 0.10) = 50(1.10)$; at the end of the second year, the $50(1.10)$ is increased by 10% of itself to become $50(1.10)(1.10)$, or $50(1.10)^2$; at the end of the third year, the $50(1.10)^2$ is increased by 10% of itself to become $50(1.10)^2(1.10)$, or

$50(1.10)^3$. If y denotes this investment's value after x years, then y and x are related by the equation $y = 50(1.10)^x$.

(a) Use a spreadsheet to create a table that gives the value of this investment for each of the following x-values: 0, 1, 2, 3, 4, 5, 6, 7, 8, 9, 10.
(b) Use Chart Wizard to create the corresponding graph.
(c) Give the investment's value after 7 years.
(d) Give the investment's value after 9 years.

16. *Exponential growth: Compound interest.* An investment of $30 earns interest at 12% compounded annually.

(a) If y denotes this investment's value after x years, then write the exponential equation that expresses the relationship between y and x.
(b) Use a spreadsheet to create a table that gives the value of this investment for the x-values 0, 1, 2, 3, 4, 5, 6, 7, 8, 9, 10.
(c) Use Chart Wizard to create the corresponding graph.
(d) Give the investment's value after 5 years.
(e) Give the investment's value after 8 years.

17. *Exponential growth: Rate of increase.* The earnings per share (EPS) of a company has an initial value of $5 and increases by 30% per year (i.e., compounded annually) so that the EPS's value, y, after x years is given by $y = 5 \cdot 1.3^x$. The earnings per share (EPS) of another company also has an initial value of $5 but it increases by 60% per year (i.e., compounded annually) so that its value, y, after x years is given by $y = 5 \cdot 1.6^x$.

(a) Create a table of x- and y-values for the equations $y = 5 \cdot 1.3^x$ and $y = 5 \cdot 1.6^x$. Use the x-values 0, 1, 2, 3, 4, 5, 6.
(b) Use Chart Wizard to create the corresponding graphs.
(c) Observe the table to determine the y-intercepts and verify this result by checking the graph.
(d) Write the equation of the steeper graph.
(e) Does the steeper graph result from the higher interest rate? Clearly, the rate of increase is the engine that drives exponential growth.

18. *Identifying exponential growth or decay.* Often we are presented with data in table form such as that in Spreadsheet 9-5. Here, the column of y-values gives the earnings per share of a company for successive time periods, denoted by x.

SPREADSHEET 9-5

X	y	RATIO
0	2	
1	3	1.5
2	4.5	1.5
3	6.75	1.5

Earlier in this section, we stated that a distinctive property of exponential growth is that *each successive y-value is the same multiple of its preceding y-value and that this multiple* is b, the *base* of the exponential equation $y = ab^x$. The multiple, b, is given in the column labeled RATIO in Spreadsheet 9-5 and is determined by entering the formula **=B3/B2** in cell C3 and *copying* the formula through the remainder of the column as illustrated in Spreadsheet 9-5. It should be noted that *real-world* data are considered to exhibit exponential growth if the RATIO column contains multiples that are close to each other, if not exactly the same. We note that all of these remarks are true for *exponential decay*, but the *multiple is less than 1*.

(a) Use a spreadsheet to verify the RATIO column of Spreadsheet 9-5.
(b) For each set of data in Spreadsheet 9-6, use a spreadsheet to determine whether the data exhibit exponential growth or decay. If the data exhibit exponential growth, state so; if the data exhibit exponential decay, state so.

SPREADSHEET 9-6

DATA SET 1			DATA SET 2			DATA SET 3	
x	y		x	y		x	y
0	4		0	10		0	6
1	10		1	8		1	7.2
2	25		2	6.4		2	8.86
3	62.5		3	5.12		3	10.5
4	156.25		4	4.096		4	12.7

19. Use Excel to create and complete the table shown in the spreadsheet below. Use the result to show that $e^h \approx 1 + h$ for small values of h.

h	e^h				
0.05					
0.04			To show that		
0.03					
0.02			$e^h \approx 1 + h$		
0.01					
0.005					
0.001			for small values of h		
0.0001					
0.00001					

Note: Enter the formula **=EXP(A2)** in cell B2 and copy the formula down through the remaining cells in Column B as indicated in the above spreadsheet.

20. Use Excel to create and complete the tables shown in the following spreadsheet. Use the results to estimate the following limit (if it exists). $\lim_{h \to 0} \dfrac{e^h - 1}{h}$

h	(e^h-1)/h		h	(e^h-1)/h
0.05			-0.05	
0.04			-0.04	
0.03			-0.03	
0.02			-0.02	
0.01			-0.01	
0.005			-0.005	
0.001			-0.001	
0.0001			-0.0001	
0.00001			-0.00001	

Note: Enter the formula =(EXP(A2)-1)/A2 in cell B2 to express $\dfrac{e^h - 1}{h}$ and copy the formula down through the remaining cells of the column in the above spreadsheet. Repeat the process for the column corresponding to the negative values of h in the above spreadsheet.

21. *Exponential versus polynomial growth.* The following spreadsheet gives a graph, along with tables of x- and y-values, for $y = e^x$ and $y = x^4$ for $x > 0$. Note that the y2/y1-values are obtained by entering the formula =C2/B2 in cell D2. Also, note that the graph used only the x-values between $x = 0$ and $x = 10$ so that the resulting scale produces a useful graph.

Study the graph and tables in the spreadsheet and answer the following.
(a) Write the equation of the top graph on the interval $0 < x < 1$.
(b) Compare the y-values of the two equations for $x = 1$ and $x = 2$. Explain why we can conclude that the graphs intersect for some x- value between $x = 1$ and $x = 2$.

(c) The column labeled y2/y1 gives the ratio $\dfrac{x^4}{e^x}$. Explain why a ratio greater the 1 indicates that the polynomial function, $y = x^4$, gives the top graph, whereas a ratio less than 1 indicates that the exponential function is the top graph. Furthermore, explain how we can use this ratio to signal crossover, or intersection points, of the two graphs.
(d) Compare the y-values of the two equations for $x = 8$ and $x = 9$. Explain why we can conclude that the graphs intersect for some x-value between $x = 8$ and $x = 9$.
(e) Use the ratio y2/y1 to explain why we can conclude that the graphs intersect for some x-value between $x = 8$ and $x = 9$.
(f) Write the equation of the top graph on the interval $x > 9$.

Summary
This example provides one instance demonstrating that for larger x-values, exponential growth outpaces polynomial growth. This result is more formally stated in terms of limits as

$$\lim_{x \to \infty} \frac{x^p}{e^x} = 0 \text{ for } p > 0.$$

170

(g) Explain why the $y2/y1$ ratios in the spreadsheet support the above summary for the case $p = 4$.

x	y1 = e^x	y2 = x^4	y2/y1					
0	1	0	0					
1	2.718282	1	0.367879					
2	7.389056	16	2.165365					
3	20.08554	81	4.032753					
4	54.59815	256	4.688804					
5	148.4132	625	4.211217					
6	403.4288	1296	3.212463					
7	1096.633	2401	2.189429					
8	2980.958	4096	1.374055					
9	8103.084	6561	0.809692					
10	22026.47	10000	0.453999					
20	4.85E+08	160000	0.00033					
50	5.18E+21	6250000	1.21E-15					
100	2.69E+43	1E+08	3.72E-36					

y=e^x vs y=x^4

22. *Exponential versus polynomial growth.* Repeat Exercise 21 using the functions $y = e^x$ and $y = x^6$.

(a) Use the x-values 0, 1, 5, 10, 11, 12, 13, 14, 15, 16, 17, 18, 19, 20, 50, 100, and 200. You do not have to create the accompanying graph.

(b) Compare the y-values of both equations for $x = 1$ and $x = 5$. Explain why we can conclude that the graphs intersect for some x-value between $x = 1$ and $x = 5$.

171

(c) Use the ratio $y2/y1$ to explain why we can conclude that the graphs intersect for some x-value between $x = 1$ and $x = 5$.

(d) Use the ratio $y2/y1$ to explain why we can conclude that the graphs intersect for some x-value between $x = 16$ and $x = 17$.

(e) Write the equation of the top graph on the interval $x > 17$.

(f) Explain why, for the case $p = 6$, the $y2/y1$ ratios support the summary given at the end of Exercise 21.

(g) This example provides another instance demonstrating that for larger x-values, _____ growth outpaces _____ growth.

9-2 The Constant e and Continuous Compounding

In Exercise 5 of Section 8-1, we learned that the constant e is defined as

$$e = \lim_{x \to \infty}\left(1 + \frac{1}{x}\right)^x \quad \text{where } e = 2.7182818\ldots.$$

In other words, the form $\left(1 + \frac{1}{x}\right)^x$ approaches e as $x \to \infty$. This is indicated in Spreadsheet 9-11.

SPREADSHEET 9-11

x	(1+1/x)^x
1	2
10	2.59374246
100	2.704813829
1000	2.716923932
10000	2.718145927
100000	2.718268237
1000000	2.718280469
10000000	2.718281694
100000000	2.718281786

Now, we show how the number e occurs naturally in the application of compound interest. As we will learn in homework Exercise 1, for a given annual rate, the more frequently interest is compounded, the greater the growth of the investment. Specifically, let's consider the growth of a $1000 investment at 6% compounded annually, semiannually, quarterly, monthly, and daily (assuming 360 days per year). The value of the investment after x years is given by the following formulas.

<u>Compounded annually</u>   (x years): $y = 1000(1.06)^x$

<u>Compounded semiannually</u>   ($2x$ half-years)   : $y = 1000(1.03)^{2x}$

<u>Compounded quarterly</u>   ($4x$ quarters): $y = 1000\left[1 + \dfrac{0.06}{4}\right]^{4x}$

<u>Compounded monthly</u>   (12x months) : $y = 1000\left[1 + \dfrac{0.06}{12}\right]^{12x}$

<u>Compounded daily</u>   (360x days): $y = 1000\left[1 + \dfrac{0.06}{360}\right]^{360x}$

<u>Compounded continuously</u>

Let m denote the number of compoundings per year. Then, if our \$1000 investment grows at 6% compounded m times per year, the formula for the investment's value after x years is

$$y = 1000\left[1 + \dfrac{0.06}{m}\right]^{mx}$$

where mx denotes the number of compoundings during x years. If interest is compounded continuously, then m increases without bound. The preceding formula is rewritten (using laws of exponents) equivalently as

$$y = 1000\left[\left(1 + \dfrac{1}{m/0.06}\right)^{(m/0.06)}\right]^{0.06x} \qquad (1)$$

As m increases without bound, $m/0.06$ also increases without bound, so the expression inside the square brackets is of the form

$$\left(1 + \dfrac{1}{k}\right)^{k}$$

with $k = m/0.06$. Thus k increases without bound (i.e., $k \to \infty$), so the expression inside the square brackets approaches e as $k = m/0.06$ approaches ∞. Equation (1) becomes

$$y = 1000e^{0.06x},$$

which gives the value of \$1000 compounded continuously at 6% after x years.

Thus, after 5 years, the original \$1000 has increased to

$$y = 1000e^{0.06(5)}$$
$$= 1000e^{0.30} = \$1349.86.$$

> **Continuous Compounding**
>
> If an initial amount, P, grows at r% compounded continuously for x years, then its accumulated (compound) amount after x years is given by
> $$y = Pe^{rx}$$

EXERCISES

1. Frequency of compounding. Spreadsheet 9-12 illustrates the growth of the $1000 investment (illustrated in this section) at 6% compounded annually, semiannually, quarterly, monthly, and daily (assuming 360 days per year) and continuously during a 5- year time interval.

SPREADSHEET 9-12

Frequency of compounding	m	Principal	(1+0.06/b2)^(b2*5)	CompAmt
Annually	1	1000	1.338225578	1338.23
Semiannually	2	1000	1.343916379	1343.92
Quarterly	4	1000	1.346855007	1346.86
Monthly	12	1000	1.348850153	1348.85
Daily	360	1000	1.349825065	1349.83
Continuously		1000	1.349858808	1349.86

Recall that m denotes the number of compoundings per year. Cell D2 contains the formula =(1+0.06/B2)^(B2*5), which was copied down through cell D6. These are the compound amount factors $\left(1 + \dfrac{0.06}{m}\right)^{mx}$. Cell D8 contains the formula =EXP(0.06*5), which is the compound amount factor for continuous compounding. Of course, the compound amount column is the product of Columns C and D.

(a) Copy Spreadsheet 9-12 for an annual rate of 8%. State what happens to the compound amount as the frequency of compounding increases.
(b) Copy Spreadsheet 9-12 for an annual rate of 6% and a time period of 7 years. State what happens to the compound amount as the frequency of compounding increases.
(c) Copy Spreadsheet 9-12 for an annual rate of 9% and a time period of 8 years. State what happens to the compound amount as the frequency of compounding increases.

2. ***Continuous compounding.*** Use a spreadsheet to compute the accumulated (compound) amount of each of the following investments assuming continuous compounding of interest at the indicated annual rates.

	PRINCIPAL	ANNUAL RATE	TIME
(a)	$5000	5%	3 years
(b)	$8000	6%	8 years
(c)	$4000	7%	6 years
(d)	$7000	8%	5 years

3. ***Continuous compounding:*** <u>Pencil and Paper Exercise</u>. Use a calculator to confirm the accumulated (compound) amount of each investment in Exercise 2.

4. ***Continuous compounding: Graphical display.*** Nine thousand dollars is invested at 7% compounded continuously.
(a) Write the equation that gives the accumulated (compound) amount after x years.
(b) Use the x-values 0, 1, 2, 3, 4, 5, 6, 7, and 8 to create a table of x- and y-values for the equation of part (a).
(c) Use Chart Wizard to create the corresponding graph.
(d) Identify the y-intercept.

5. ***Continuous compounding: Graphical display.*** Two $5000 investments are made. One earns interest at 4% compounded continuously and the other at 9% compounded continuously.
(a) Write the equation that gives the accumulated (compound) amount after x years for each investment.
(b) Use the x-values 0, 1, 2, 3, 4, 5, 6, 7, and 8 to create a table of x- and y-values for the equations of part (a).
(c) Use Chart Wizard to create the corresponding graphs.
(d) Identify the y-intercept for each graph.
(e) Write the equation of the steeper graph. Does the steeper graph result from the higher interest rate?

6. ***Continuous compounding: Graphical display.*** <u>Pencil and Paper Exercise</u>. Ten thousand dollars is invested at 6% compounded continuously.
(a) Write the equation that gives the accumulated (compound) amount after x years.
(b) Draw the graph of the equation in part (a).
(c) Identify the y-intercept.

7. ***Continuous compounding: Graphical display.*** <u>Pencil and Paper Exercise</u>. Two $8000 investments are made. One earns interest at 5% compounded continuously and the other at 10% compounded continuously.
(a) Write the equation that gives the accumulated (compound) amount after x years for each investment.
(b) Draw the graphs of the equations in part (a).
(c) Identify the y-intercept for each graph.
(d) Write the equation of the steeper graph. Does the steeper graph result from the higher interest rate?

8. *Using Goal Seek to determine the principal.* Determine the principal needed to accumulate into $20,000 after 6 years at 5% compounded continuously. Refer to Spreadsheet 9-13 and the following instructions.

SPREADSHEET 9-13

Prin P	CmpAmt y
	0

INSTRUCTIONS FOR SPREADSHEET 9-13

1. Type the labels as shown.
2. In cell B2, type the compound amount (continuous compounding) formula **=A2*EXP(0.05*6),** and press **Enter.** The result **0** appears in cell B2 as shown in Spreadsheet 9-13. Note that cell A2 represents the principal, *P.*
3. If it is not already there, use the mouse *to move the dark-bordered rectangle to the cell containing the formula*—in this case cell B2, which currently contains the value **0.**
4. Select **Tools** and then choose **Goal Seek,** and a dialog box appears. Note that *the cell containing the formula*—in this case cell B2—appears in the *Set cell* text box.
5. Because it is our goal to set cell B2 equal to a value of 20,000 by changing the value of cell A2, we type **20000** in the *To value* text box and **A2** in the *By changing cell* text box, and click **OK.** The required principal, P (in this case, 14,816.36) appears in cell A2. Thus, the required principal and the solution to our problem $14,816.36.

(a) Follow these instructions on your spreadsheet to verify the above results.
(b) Use *Goal Seek* to determine the principal needed to accumulate $30,000 after 7 years at 6% compounded continuously.
(c) Usc *Goal Seek* to determine the principal needed to accumulate $50,000 after 4 years at 8% compounded continuously.
(d) Use *Goal Seek* to determine the principal needed to accumulate $25,000 after 3 years at 9% compounded continuously
(e) Use *Goal Seek* to determine the principal needed to accumulate $80,000 after 9 years at 5% compounded continuously.

9. *Using Goal Seek to determine the annual rate.* Determine the annual rate, *r*, needed to grow $30,000 into $70,000 during a time interval of 4 years. Assume continuous compounding.
Hint: Reserve a cell, say cell A2, for the unknown variable, *r*. In cell B2, type the compound amount (continuous compounding) formula **=30000*EXP(A2*4)** and press **Enter**, and the result 0 appears in cell B2. Note that cell A2 represents the annual rate, *r*. Follow the remaining instructions for using *Goal Seek* in Exercise 8.

10. *Using Goal Seek to determine the annual rate.* Determine the annual rate, *r*, needed to grow $20,000 into $90,000 during a time interval of 10 years. Assume continuous compounding.

11. ***Using Goal Seek to determine the annual rate.*** Determine the annual rate, r, needed to grow $10,000 into $60,000 during a time interval of 8 years. Assume continuous compounding.

12. ***Using Goal Seek to determine growth time.*** Determine the time, t, needed to grow $15,000 into $60,000 at 5% compounded continuously.

13. ***Using Goal Seek to determine growth time.*** Determine the time, t, needed to grow $25,000 into $75,000 at 8% compounded continuously.

14. ***Using Goal Seek to determine growth time.*** Determine the time, t, needed to grow $35,000 into $56,000 at 7% compounded continuously.

15. ***Using Goal Seek to determine doubling time.*** Determine the time, t, needed to double an initial investment of $40,000 at 6% compounded continuously.
Note: There is more than one approach to this type of problem.
(1) One way is to follow the approach used in Exercises 12 through 14. Specifically, if the principal is $40,000, the compound amount is double that amount, or $80,000. Now continue the approach used in Exercises 12 through 14.

(2) Another approach involves beginning with the equation $y = Pe^{rt}$ and recognizing that doubling P implies that $e^{0.06t} = 2$. Thus, we use *Goal Seek* to solve the equation for t.

16. ***Using Goal Seek to determine doubling time.*** Determine the time, t, needed to double an initial investment at 8% compounded continuously. Note: Use the second method outlined in Exercise 15.

17. ***Using Goal Seek to determine doubling time.*** Determine the time, t, needed to double an initial investment at 20% compounded continuously. Note: Use the second method outlined in Exercise 15.

18. ***Using Goal Seek to determine tripling time.*** Determine the time, t, needed to triple an initial investment at 8% compounded continuously. Note: Use the second method outlined in Exercise 15.

19. ***Using Goal Seek to determine tripling time.*** Determine the time, t, needed to triple an initial investment at 10% compounded continuously. Note: Use the second method outlined in Exercise 15.

20. ***Using Goal Seek to determine tripling time.*** Determine the time, t, needed to triple an initial investment at 5% compounded continuously.

9-3 Logarithmic Functions

Solving for an Exponent

Consider solving the exponential equation $e^x = 8$ for x. We use *Goal Seek* to solve for x by following the instructions presented after Spreadsheet 9-14.

SPREADSHEET 9-14

x	EXP(x)
	1

INSTRUCTIONS FOR SPREADSHEET 9-14

1. Type the labels as shown.
2. Reserve cell A2 for the unknown variable, x, by leaving it blank. Then, in cell B2, type the formula =**EXP(A2)** and, press **Enter**, and the result 1 appears in cell B2 as shown in Spreadsheet 9-14. Because cell A2 is empty, Excel returns e^0, or 1 in cell B2.
3. If it is not already there, use the *mouse to move the dark-bordered rectangle to the cell containing the formula*—in this case cell B2—which currently contains the value 1.
4. Select **Tools** and then choose *Goal Seek*, and a dialog box appears. Note that *the cell containing the formula*—in this case, cell B2—appears in the *Set cell* text box.
5. Because it is our goal to set cell B2 equal to a value of 8 by changing the value of cell A2, we type **8** in the *To value* text box and **A2** in the *By changing cell* text box, and click **OK**. The solution value of x, 2.079383, appears in cell A2, and 7.999533 appears in cell B2 as the closest approximation to e^x, or 8.

Thus, the solution to the exponential equation $e^x = 8$ is $x = 2.079383$. The solution is also called the *natural logarithm of 8* and can be written as **2.079 = ln 8**, where we rounded the solution to three decimal places.

The Meaning of a Natural Logarithm

The solution, x, to an exponential equation of the form $e^x = k$ is the natural logarithm of k and is written as $x = \ln k$. In other words, the natural logarithm of k is the exponent of e for which $e^x = k$.

Natural Logarithmic Function

With Excel, the natural logarithm of a number x is determined by using the formula =*LN(x)*. Thus, if we move our pointer to a cell and type =**LN(8)**, the number 2.079, rounded to three decimal places, appears in the cell. A graph of the equation $y = \ln x$ is created by finding the natural logarithms of the indicated x-values in Spreadsheet 9-15. The equation $y = \ln x$ defines the *natural logarithmic function*. Studying its graph in

Spreadsheet 9-15, note that the *y*-axis is a *vertical asymptote*. Also, note that the *y*-values of the natural logarithmic function increase very slowly for increasing *x*-values. Specifically, note that the *y*-value corresponding to $x = 50$ is approximately 4, whereas the *y*-value corresponding to $x = 100$ is approximately 4.6. This is not a large increase in *y*-values corresponding to a 50- unit increase in *x*-values.

SPREADSHEET 9-15

x	y = ln x
0.1	-2.302585
0.3	-1.203973
0.5	-0.693147
0.9	-0.105361
1	0
5	1.609438
10	2.302585
30	3.401197
50	3.912023
80	4.382027
100	4.60517
150	5.010635
200	5.298317

INSTRUCTIONS FOR SPREADSHEET 9-15

Type the formula =LN(A2) in cell B2 and copy the formula through Column B as illustrated in Spreadsheet 9-15.

EXERCISES

1 – 5. *Solving for an exponent.* Each of the following exponential equations is of the form $e^x = k$.
  (a) Use a spreadsheet to solve each of the following for *x*.
  (b) Use a spreadsheet to verify that the solution, *x*, equals ln *k*. In other words, use the formula =LN(*k*) to verify that the formula result equals the solution value of *x*.

1. $e^x = 5$ 2. $e^x = 15$ 3. $e^x = 40$ 4. $e^x = 65$ 5. $e^x = 90$

6 – 10. *Rewriting exponential equations in logarithmic form.* Without using its solution, rewrite each equation of Exercises 1 through 5 in logarithmic form, $x = \ln k$.

11. *Rewriting an exponential equation in logarithmic form.* Rewrite the equation $y = e^x$ in logarithmic form.

12. *Rewriting a logarithmic equation in exponential form.* Rewrite the equation $y = \ln x$ in exponential form.

13. *Understanding the meaning of a natural logarithm.* Return to the natural logarithmic function in Spreadsheet 9-15. Consider a point—for example, (50, 3.912023)— on the natural logarithmic function $y = \ln x$, and verify that $x = e^y$. In other words, for the point (50, 3.912023), use the formula =**EXP(3.912023)** to verify that $50 = e^{3.912023}$.
(a) Repeat the above for the point (80, 4.382027).
(b) Repeat the above for the point (0.1, -2.30259).
(c) Repeat the above for the point (0.5, -0.69315).
(d) Repeat the above for the point (100, 4.60517).
(e) Repeat the above for the x-intercept, (1, 0).

14. *Properties of the natural logarithmic function.* Return to the natural logarithmic function in Spreadsheet 9-15.
(a) Verify that the x-intercept is (1, 0).
(b) On what interval of x-values is the natural logarithmic function negative?
(c) On what interval of x-values is the natural logarithmic function positive?
(d) Verify that the natural logarithmic function increases slowly by computing the change in y-values as x changes from 100 to 150.
(e) Verify that the natural logarithmic function increases slowly by computing the change in y-values as x changes from 150 to 200.

15. *Graphs of $y = e^x$ and $y = \ln x$.*
(a) Use the x-values -2, -1, 0, 1, 2, 3, 4, and 5 to create a table of x- and y-values, along with the corresponding graph, $y = e^x$.
(b) Keeping the results of part (a) intact in the spreadsheet, move to a location in the spreadsheet below the results of part (a) and create a new table of x- and y-values by interchanging the x- and y-columns of part (a). Next, use Chart Wizard to create the corresponding graph.
(c) Explain why part (b) gives the graph of the natural logarithmic function, $y = \ln x$.
(d) Explain how to get the graph of $y = \ln x$ from that of $y = e^x$, and vice versa.

16. *Logarithmic versus polynomial growth.* The following spreadsheet gives a graph, along with tables of x- and y-values, for $y = \ln x$ and $y = x^{1/2}$ for $x > 0$. Observe that we have made no entry in the ln x column for $x = 0$ because ln 0 is undefined. Therefore, the formula =**LN(A3)** was typed in cell B3 and copied down through the column.

Additionally, the formula =A2^(1/2) was typed in cell B2 and copied down through the column. The $y1/y2$-values were obtained by entering the formula =B3/C3 in cell D3 and were copied down through the column.

x	y1 = ln x	y2=x^(1/2)	y1/y2					
0		0						
0.5	-0.693147	0.707107	-0.980258					
1	0	1	0					
2	0.693147	1.414214	0.490129					
5	1.609438	2.236068	0.719763					
10	2.302585	3.162278	0.728141					
20	2.995732	4.472136	0.669866					
30	3.401197	5.477226	0.620971					
50	3.912023	7.071068	0.553244					
100	4.60517	10	0.460517					
200	5.298317	14.14214	0.374648					
500	6.214608	22.36068	0.277926					
1000	6.907755	31.62278	0.218442					
10000	9.21034	100	0.092103					

y = ln x vs y = x^(1/2)

When creating the graph, we used only the x-values between $x = 0$ and $x = 50$ so that the resulting scale would produce a useful graph. Thus, we shaded only the first three columns, including the labels down through $x = 50$. Finally, note that these graphs are similar, which is why we compare them.

Study the graph and tables in the spreadsheet and answer the following.
  (a) Write the equation of the bottom graph on the interval $x > 0$.

  (b) The column labeled $y1/y2$ gives the ratio $\dfrac{\ln x}{x^{1/2}}$. Explain why a ratio less than 1 indicates that $y = \ln x$ gives the bottom graph, whereas a ratio greater than 1 indicates the opposite.
  (c) Using the $y1/y2$ ratios, write the equation of the bottom graph on the interval $x > 0$.

Summary
This example provides one instance demonstrating that for larger x-values, polynomial growth outpaces logarithmic growth. In other words, for larger x-values, logarithmic growth is slower than polynomial growth. This result is more formally stated in terms of limits as $\lim\limits_{x \to \infty} \dfrac{\ln x}{x^{p}} = 0$ for $p > 0$.

  (d) Explain why the $y1/y2$ ratios in the spreadsheet support the above summary for the case $p = 1/2$.

17. *Logarithmic versus polynomial growth.* Repeat Exercise 16 using the functions $y = \ln x$ and $y = x^{1/3}$.
  (a) Use the x-values 0, 0.5, 2, 5, 10, 20, 30, 50, 100, 200, 500, 1000, 10,000, 50,000, 100,000, 1,000,000, 10,000,000 and 100,000,000. Follow the procedure used in Exercise 16 to avoid ln 0.
  (b) Create the accompanying graph, using only the x-values from 0 to 1000.
  (c) Compare the y-values of both equations for $x = 5$ and $x = 10$. Explain why we can conclude that the graphs intersect for some x-value between $x = 5$ and $x = 10$.
  (d) Use the ratio $y1/y2$ to also explain why we can conclude that the graphs intersect for some x-value between $x = 5$ and $x = 10$.
  (e) Use the ratio $y1/y2$ to also explain why we can conclude that the graphs intersect for some x-value between $x = 50$ and $x = 100$.
  (f) Write the equation of the bottom graph on the interval $x > 100$.
  (g) Explain why, for the case $p = 1/3$, the $y1/y2$ ratios support the summary given at the end of Exercise 16.
  (h) This example provides another instance demonstrating that for larger x-values, _____ growth outpaces _____ growth or, equivalently, _____ growth is slower than _____ growth.

18. *Graph of ln y versus x for exponential growth or decay.* The top portion of Spreadsheet 9-16 contains a table and graph for the exponential function $y = 5 \cdot 3^{x}$. The bottom portion of Spreadsheet 9-16 contains a table and graph of the ln y-values versus x-

values. In other words, the ln y-values are the natural logs of the y-values of the exponential function $y = 5 \cdot 3^x$.

SPREADSHEET 9-16

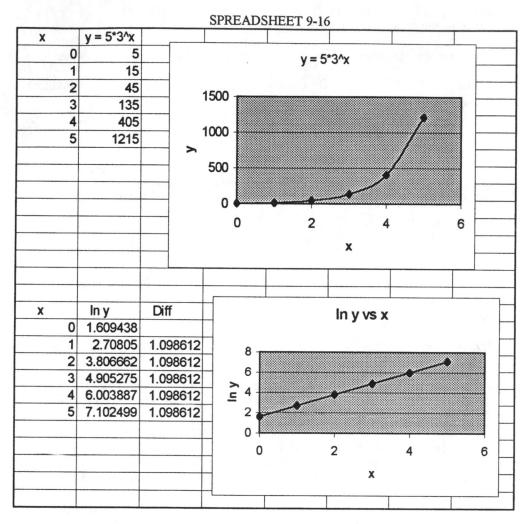

Observe that the graph of the ln y-values versus x-values is a straight line. The *slope* of the straight line is 1.098612 as given in the "Diff" column next to the ln y-values. The "Diff" column was determined by subtracting successive ln y-values.

Note that as we move down the y-column of the exponential function $y = 5 \cdot 3^x$, each successive y-value is the *same multiple* of (in this case, 3 times) the preceding y-value. As mentioned in Exercise 18 of Section 9-1, this is a distinctive property of exponential growth or decay. It is equivalent to saying that *the percent change in the y-values is constant*. This *constant percent change* in y-values translates into a *constant slope* (i.e., a straight line) when we graph ln y-values versus x-values.

This example illustrates the following result, which is true of exponential functions.

If a graph of y-values versus x-values exhibits exponential growth or decay, the corresponding graph of ln y-values versus x-values has the graph of a straight line.

This is true because the exponential form

$$y = ab^x$$

is equivalent to the logarithmic form

$$\ln y = \ln a + (\ln b)x$$

which shows that $\ln y$ and x are linearly related.

(a) <u>Pencil and Paper Exercise</u>. Compare $y = 5 \cdot 3^x$ to the exponential form $y = ab^x$ and state a and b.

(b) <u>Pencil and Paper Exercise</u>. Study the bottom portion of Spreadsheet 9-16 and verify that $\ln b$ is the *slope* of the straight line that gives the relationship between $\ln y$ and x.

(c) <u>Pencil and Paper Exercise</u>. Study the bottom portion of Spreadsheet 9-16 and verify that $\ln a$ is the *y-intercept* of the straight line that gives the relationship between $\ln y$ and x.

19. *Graph of ln y versus x for exponential growth or decay.* Repeat Exercise 18 by creating a table and graph for the exponential function $y = 2 \cdot 5^x$, using the x-values 0, 1, 2, 3, 4, and 5. Then, create a table and graph for the corresponding $\ln y$- versus x-values.

20. *Graph of ln y versus x for exponential growth or decay.* Use properties of logarithms to show that the exponential form $y = ab^x$ is equivalent to the logarithmic form $\ln y = \ln a + (\ln b)x$.

21. *Graph of ln y versus x for exponential growth or decay.*
Summary. If a table of y-values exhibits exponential growth or decay, then the *percent change* in y-values across equally spaced x-values is _____.   Furthermore, a graph of $\ln y$-values versus x-values results in a _____ line.

22. *Graph of ln y versus x for exponential growth or decay.*
Summary. If a table of y-values exhibits exponential growth or decay, then the *constant percent change* in y-values across equally spaced x-values translates into a _____ _____ when we graph $\ln y$-values versus x-values.

CHAPTER TEN

Y', Y'', Acceleration, and Graphical Analysis

10-1 Tables of Y'- and Y''-Values; Acceleration

Given a function $y = f(x)$, we have already learned that the first derivative, y', gives the *instantaneous change in y per unit change in x*, and the second derivative, y'', gives the *instantaneous change in y' per unit change in x*. Sometimes these concepts can be seen more clearly by referring to tables of x- and y-values.

The table in Spreadsheet 10-1 gives a company's earnings per share (EPS) for a succession of years. Investors often study such data to analyze the movement of earnings growth. Here, x denotes time in years and y denotes the earnings per share (EPS).

SPREADSHEET 10-1

X	Y	Y'	Y''
1	10		
2	50	40	
3	15	-35	-75
4	40	25	60
5	90	50	25

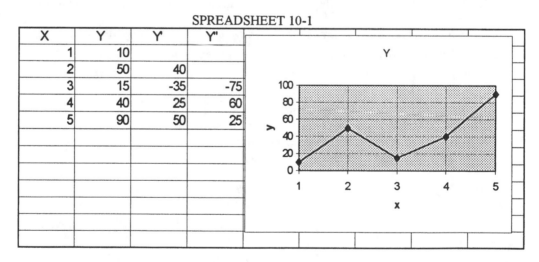

Y'-*Values*

The y'-values are determined by subtracting successive y-values as we move down the table. In other words, the y'-*values* give the *change in y-values per unit change in x* and are *approximate* values of the *derivative* because they are slopes of the straight-line segments connecting the individual points of the graph given in Spreadsheet 10-1.

186

Specifically, observe that as we move down the table from $x = 1$ to $x = 2$, the y-value changes from $y = 10$ to $y = 50$. The *change in y,* 40, is given in the y'-column. Note also that as we move down the table from $x = 2$ to $x = 3$, the y-values change from $y = 50$ to $y = 15$. Here the change in y is a decrease of 35, which is appropriately given in the y' column as $y' = -35$.

Y''-Values

The y''-values are determined by subtracting successive y'-values as we move down the table. In other words, the y''-*values* give the *change in y'-values per unit change in x* and are approximate values of the second derivative. Specifically, observe that as we move down the table from $x = 2$ to $x = 3$, the y'-value changes from $y' = 40$ to $y' = 35$. The *change in y',* -75, is given in the y'' column. Note also that as we move down the table from $x = 3$ to $x = 4$, the y'-values change from $y' = -35$ to $y' = 25$. Here the change in y' is an increase of 60, which is appropriately given in the y'' column as $y'' = 60$.

Acceleration and Deceleration. Observe that *positive y''-values correspond to y'-values that are increasing, and thus the corresponding y-values are* **accelerating**, whereas **negative y''-values** *correspond to y'-values that are decreasing, and thus the corresponding y-values are* **decelerating**.

Thus, as we move from $x = 2$ to $x = 3$, the *negative y''-value*, $y'' = -75$, indicates that the y-values (EPS) are *decelerating*. As we move from $x = 3$ to $x = 4$, the *positive y''-value*, $y'' = 60$, indicates that the y-values (EPS) are *accelerating*.

INSTRUCTIONS

Use the following instructions to create tables of y'- and y''-values as in Spreadsheet 10-1.

1. Assume that we've opened Excel, typed the column labels x, y, y', and y'' in cells A1, B1, C1, and D1, respectively, and entered the x- and y-values.

2. Because the y'-values are the differences between the successive y-values as we move down the table, move the dark-bordered rectangle to cell C3 of the y'-column and type **=B3-B2.** This *formula* gives the *change in y-values* as we move from the first to the second y-value.

3. Now, we *copy the formula* down through cell C6 as follows:
   Move the dark-bordered rectangle to cell C3.
   Use the mouse to move the pointer to the small black box (called a handle) at the lower right corner of cell C3. The mouse pointer becomes a thick black plus.
   Click the mouse button without releasing it and drag the mouse pointer down to cell C6.
   Release the mouse button at cell C6, and cells C3 through C6 will contain the formula values.

4. Repeat this process to create the y''-values in Column D. Remember that the y''-values are the differences in successive y'-values as we move down the table.

EXERCISES

For each of the tables in Spreadsheet 10-2 below, x denotes time in years and y denotes EPS for some company.

SPREADSHEET 10-2

TABLE 1				TABLE 2			
X	Y	Y'	Y"	X	Y	Y'	Y"
1	10			1	80		
2	15			2	70		
3	24			3	56		
4	46			4	29		
5	53			5	20		
6	57			6	18		

1. Refer to Table 1 in Spreadsheet 10-2.
   (a) Fill in the y' and y'' columns.
   (b) State the y'-value corresponding to $x = 2$ and give its meaning.
   (c) State the y'-value corresponding to $x = 4$ and give its meaning.
   (d) State the y''-value corresponding to $x = 4$ and give its meaning.
   (e) Is EPS accelerating or decelerating as we move from $x = 3$ to $x = 4$? Explain why.
   (f) Is EPS accelerating or decelerating as we move from $x = 4$ to $x = 5$? Explain why.
   (g) Explain why some of the y''-values are negative despite the fact that the y-values are increasing.

2. Refer to Table 2 in Spreadsheet 10-2.
   (a) Fill in the y' and y'' columns.
   (b) State the y'-value corresponding to $x = 2$ and give its meaning.
   (c) State the y'-value corresponding to $x = 3$ and give its meaning.
   (d) State the y''-value corresponding to $x = 3$ and give its meaning.
   (e) Is EPS accelerating or decelerating as we move from $x = 3$ to $x = 4$? Explain why.
   (f) Is EPS accelerating or decelerating as we move from $x = 4$ to $x = 5$? Explain why.
   (g) Explain why some of the y''-values are positive despite the fact that the y-values are decreasing.

3. Pencil and Paper Exercise. *(Significance of the sign of y')*
A *positive y'-value* indicates that the corresponding y-values are _____.
A *negative y'-value* indicates that the corresponding y-values are _____.

4. Pencil and Paper Exercise. *(Significance of the sign of y'')*
A *positive y''-value* indicates that the corresponding y'-values are _____ and the y-values are _____.
A *negative y''-value* indicates that the corresponding y'-values are _____ and the y-values are _____.

188

SPREADSHEET 10-3

	TABLE 1				TABLE 2		
X	Y	Y'	Y''	X	Y	Y'	Y''
1	8			1	10		
2	3			2	35		
3	7			3	30		
4	13			4	19		
5	24			5	15		
6	25			6	14		

5. <u>Pencil and Paper Exercise</u>. Refer to Table 1 in Spreadsheet 10-3.
   (a) Fill in the y' and y'' columns.
   (b) State the y'-value corresponding to $x = 2$ and give its meaning.
   (c) State the y'-value corresponding to $x = 4$ and give its meaning.
   (d) State the y''-value corresponding to $x = 4$ and give its meaning.
   (e) Are the y-values accelerating or decelerating as we move from $x = 3$ to $x = 4$?
   Explain why.
   (f) Are the y-values accelerating or decelerating as we move from $x = 5$ to $x = 6$?
   Explain why.
   (g) Explain why some of the y''-values are negative despite the fact that the y-values are increasing.

6. <u>Pencil and Paper Exercise</u>. Refer to Table 2 in Spreadsheet 10-3.
   (a) Fill in the y' and y'' columns.
   (b) State the y'-value corresponding to $x = 2$ and give its meaning.
   (c) State the y'-value corresponding to $x = 3$ and give its meaning.
   (d) State the y''-value corresponding to $x = 3$ and give its meaning.
   (e) Are the y-values accelerating or decelerating as we move from $x = 3$ to $x = 4$?
   Explain why.
   (f) Are the y-values accelerating or decelerating as we move from $x = 4$ to $x = 5$?
   Explain why.
   (g) Explain why some of the y''-values are positive despite the fact that the y-values are decreasing.

7. *Projectile.* A ball is projected vertically into the air. The function defined by

$$S = -16t^2 + 192t \qquad (0 \le t \le 12)$$

gives the height of the ball (in feet) above the ground at time t (in seconds).
   (a) Create a table of t-, S-, S'-, and S''-values. Use the t-values 0, 1, 2, . . . 12.
   (b) Is the ball accelerating or decelerating as it moves through the air? Explain why.
   (c) Use Chart Wizard to create the corresponding graph of S-values versus t-values.

10-2 Graphs of f, f' and f''; Concavity and Acceleration

What Does the Graph of f' Reveal about the Graph of f?

Spreadsheet 10-4 gives tables and graphs for $f(x) = x^2 + 5$ and its derivative $f'(x) = 2x$.

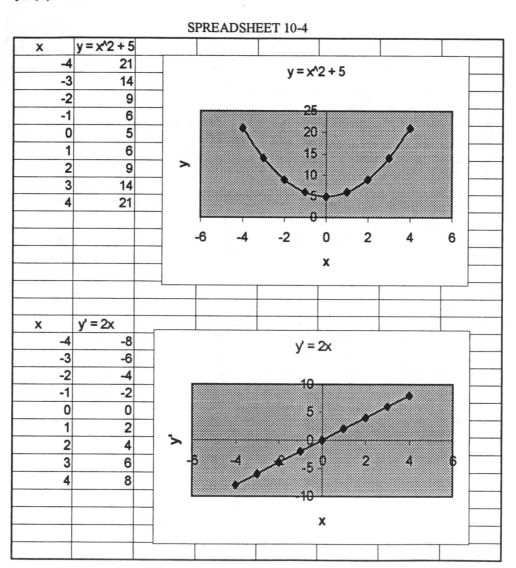

SPREADSHEET 10-4

Studying the graph of $f(x) = x^2 + 5$ in Spreadsheet 10-4, note that f is decreasing on the interval $x < 0$ and increasing on $x > 0$. Accordingly, the graph of its derivative, $f'(x) = 2x$, is negative on the interval $x < 0$ and positive on $x > 0$. Also, note that $f' = 0$ at $x = 0$ to indicate that $x = 0$ is a *critical value* and, therefore, a candidate for a local maximum or minimum. The fact that $f' = 0$ at $x = 0$ indicates that f has a horizontal tangent line at $x = 0$. Remember that a *critical value* of f is a value $x = p$ in the domain of f where $f'(p) = 0$ or $f'(p)$ is undefined. Thus, we summarize as follows.

Where f is increasing or decreasing:
If f' is *positive* on an interval, then f *is increasing* on that interval.
If f' is *negative* on an interval, then f *is decreasing* on that interval.
If $f' = 0$ at a point, then f *is neither increasing nor decreasing* but has a *horizontal tangent* line at that point.

We also use the first derivative to confirm the existence of a local maximum or minimum at a critical value by looking for a sign change in the first derivative at a critical value. Again referring to Spreadsheet 10-4, observe that f' *changes from negative to positive* at the critical point $x = 0$ as we move along its graph in a left-to-right direction across $x = 0$. This confirms the existence of a local minimum at $x = 0$, which we verify by observing the graph of f.

Where f has local maxima or minima:
Assume that f' exists on both sides of a critical value $x = p$.
If f' *changes from negative to positive* at a critical value $x = p$ as we move along its graph in a left-to-right direction, then f has a *local minimum* at the critical value $x = p$.
If f' *changes from positive to negative* at a critical value $x = p$ as we move along its graph in a left-to-right direction, then f has a *local maximum* at the critical value $x = p$.

What Does the Graph of f'' Reveal about the Graph of f?
Spreadsheet 10-5 gives tables and graphs for $f(x) = x^3 - 12x^2 + 36x + 10$ and its second derivative $f''(x) = 6x - 24$. Studying the graph of $f(x) = x^3 - 12x^2 + 36x + 10$ in Spreadsheet 10-5, note that f is concave down on the interval $x < 4$ and concave up on $x > 4$. Accordingly, the graph of its second derivative, $f''(x) = 6x - 24$, is negative on the interval $x < 4$ and positive on $x > 4$. The fact that the second derivative changes sign at $x = 4$ confirms the existence of an inflection point at $x = 4$. Although $f'' = 0$ at $x = 4$, we caution that the second derivative equaling 0 at a point does not necessarily indicate the existence of an inflection point. We will explore this issue in the homework exercises. Thus, we summarize as follows.

Where f is concave up or concave down:
If f'' is *positive* on an interval, then f *is concave up* on that interval.
If f'' is *negative* on an interval, then f *is concave down* on that interval.

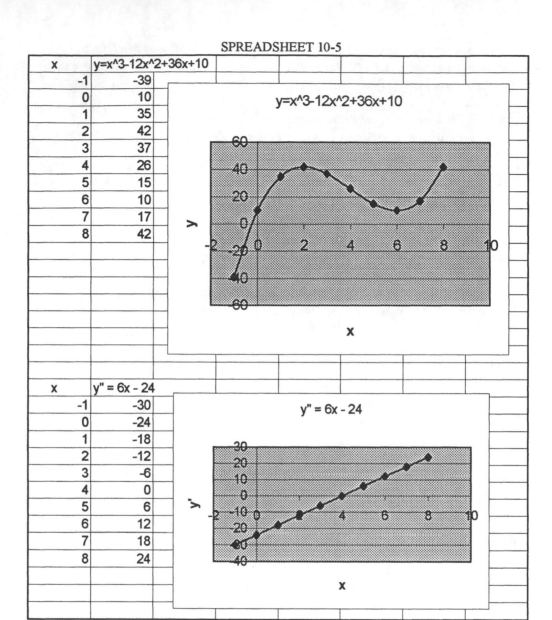

x	y=x^3-12x^2+36x+10
-1	-39
0	10
1	35
2	42
3	37
4	26
5	15
6	10
7	17
8	42

x	y" = 6x - 24
-1	-30
0	-24
1	-18
2	-12
3	-6
4	0
5	6
6	12
7	18
8	24

Concavity and Acceleration

The second derivative gives the rate of change of the first derivative. Thus, if f'' is *positive* on some interval, this indicates that f' is increasing on that interval and, therefore, the f-values (i.e., the y-values) are *accelerating* on that interval. On the other hand, if f'' is *negative* on some interval, this indicates that f' is decreasing on that interval and, therefore, the f-values (i.e., the y-values) are *decelerating* on that interval. The second derivative indicates concavity, so from a graphical perspective, acceleration is related to concavity. Specifically, if the graph of a function is *concave up* on some interval, then the y-values (i.e., the function values) are *accelerating* on that interval. On the other hand, if the graph of a function is *concave down* on some interval, then the *y*-values (i.e., the function values) are *decelerating* on that interval. Thus, we summarize as follows.

Concavity and acceleration:

If $f'' > 0$, then the graph of f is *concave up* and f is *accelerating*.

If $f'' < 0$, then the graph of f is *concave down* and f is *decelerating*.

We illustrate the above by referring to the following graph entitled *The Business Cycle*.

The Business Cycle

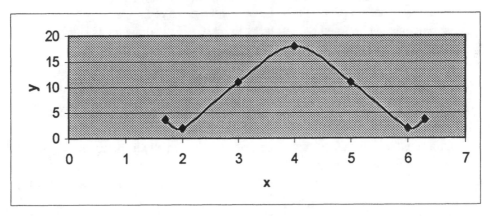

Frequently we hear, on the nightly news, that business activity is booming, or that economic growth is slow, or that a recession is imminent. All of these statements refer to the change in business activity over time. The accompanying figure gives a graph of business activity, *y*, versus time, *x*. Although it is not shown in the graph, assume that the left end-point of the graph begins at $x = 1.7$ and the right end-point ends at $x = 6.3$. The graph is *concave up* on $1.7 < x < 3$ and $5 < x < 6.3$; the graph is *concave down* on $3 < x < 5$.

The section of curve from $x = 2$ to $x = 6$ depicts a typical business cycle beginning at $x = 2$, when business activity is at a *minimum*, or *trough*, to the point at $x = 4$, when business activity is at a *maximum*, or *peak*, to the point at $x = 6$, when business activity has bottomed again. Beginning at $x = 2$ and moving rightward towards $x = 3$, business activity is both increasing and *accelerating* (*concave up*). In other words, business activity is increasing at an increasing rate. Continuing rightward towards $x = 4$, business activity is still increasing but *decelerating* (*concave down*). In other words, it is increasing at a decreasing rate. Continuing rightward towards $x = 5$, business activity is decreasing and *decelerating* (*concave down*). This means that business activity is going downhill at a faster and faster rate of decrease. At $x = 5$ the economy begins to *accelerate* (*concave up*) despite the fact that it is still decreasing. This is because its rate of decrease begins to slow down. Finally, at $x = 6$ the economy bottoms out, and the cycle begins to repeat itself.

EXERCISES

1. **Graph of f' versus f.** Create a table and graphs for $f(x) = 2x^3 - 3x^2 - 36x + 7$ and $f'(x) = 6x^2 - 6x - 36$, following the format of Spreadsheet 10-4. Use the x-values -5, -4, -3, -2, -1, 0, 1, 2, 3, 4, 5.

   (a) Observing the graph of f', state the interval(s) on which f is increasing. Confirm the results by observing the graph of f.

   (b) Observing the graph of f', state the interval(s) on which f is decreasing. Confirm the results by observing the graph of f.

   (c) Observing the graph of f', write the x-coordinates of any local maxima or minima. For each such x-coordinate, specify whether the point is a local maximum or minimum and state why. Confirm the results by observing the graph of f.

   (d) Use calculus to confirm these results by finding $f'(x)$, drawing its sign chart, and analyzing the results.

2. **Graph of f' versus f.** Create a table and graphs for $f(x) = -x^3 - 6x^2 + 180x$ and $f'(x) = -3x^2 - 12x + 180$, following the format of Spreadsheet 10-4. Use the x-values -10, -9, -8, . . . , 10.

   (a) Observing the graph of f', state the interval(s) on which f is increasing. Confirm the results by observing the graph of f.

   (b) Observing the graph of f', state the interval(s) on which f is decreasing. Confirm the results by observing the graph of f.

   (c) Observing the graph of f', write the x-coordinates of any local maxima or minima. For each such x-coordinate, specify whether the point is a local maximum or minimum and state why. Confirm the results by observing the graph of f.

   (d) Use calculus to confirm these results by finding $f'(x)$, drawing its sign chart, and analyzing the results.

3. **Graph of f' versus f.** Create a table and graphs for $f(x) = 3x^4 - 8x^3 + 20$ and $f'(x) = 12x^3 - 24x^2$, following the format of Spreadsheet 10-4. Use the x-values -1, -0.5, 0, 0.5, 1, 1.5, 2, 2.5, and 3.

   (a) Observing the graph of f', state the interval(s) on which f is increasing. Confirm the results by observing the graph of f.

   (b) Observing the graph of f', state the interval(s) on which f is decreasing. Confirm the results by observing the graph of f.

   (c) Observing the graph of f', write the x-coordinates of any local maxima or minima. For each such x-coordinate, specify whether the point is a local maximum or minimum and state why. Confirm the results by observing the graph of f.

   (d) Use calculus to confirm these results by finding $f'(x)$, drawing its sign chart, and analyzing the results.

4. *f' versus f.* Create a table and graph for $f'(x) = x^2 - 9$ using the x-values -4, -3, -2, 1, 0, 1, 2, 3, 4.

   (a) Observing the graph of f', state the interval(s) on which f is increasing.

   (b) Observing the graph of f', state the interval(s) on which f is decreasing.

   (c) Observing the graph of f', write the x-coordinates of any local maxima or minima. For each such x-coordinate, specify whether the point is a local maximum or minimum and state why.

5. *f' versus f.* Pencil and Paper Exercise. Draw the graph of $f'(x) = 4x - 12$.

   (a) Observing the graph of f', state the interval(s) on which f is increasing.

   (b) Observing the graph of f', state the interval(s) on which f is decreasing.

   (c) Observing the graph of f', write the x-coordinates of any local maxima or minima. For each such x-coordinate, specify whether the point is a local maximum or minimum and state why.

6. **Summary.** *Deriving information from a graph of f'.* Pencil and Paper Exercise. When observing a graph of f', explain how one determines:

   (a) Where f is increasing.   (b) Where f is decreasing.
   (c) Where f has a local maximum.   (d) Where f has a local minimum.

7. **When a sign change in f' does _not_ result in a local maximum or minimum.** Pencil and Paper Exercise. Draw the graph of $f(x) = \dfrac{1}{x^2}$ and its derivative $f'(x) = \dfrac{-2}{x^3}$.

   (a) Observing the graph of f', state the interval(s) on which f is increasing.

   (b) Observing the graph of f', state the interval(s) on which f is decreasing.

   (c) Observing the graph of f', state the x-coordinate of the point where f' changes sign. Explain why this point does _not_ result in a local maximum or minimum.

8. *Graph of f" versus f.* Create a table and graphs for $f(x) = 2x^3 - 3x^2 - 36x + 7$ of Exercise 1 and its second derivative $f''(x) = 12x - 6$, following the format of Spreadsheet 10-5. Use the x-values -5, -4, -3, -2, -1, 0, 1, 2, 3, 4, 5.

   (a) Observing the graph of f'', state the interval(s) on which f is concave up. Confirm the results by observing the graph of f.

   (b) Observing the graph of f'', state the interval(s) on which f is concave down. Confirm the results by observing the graph of f.

   (c) Observing the graph of f'', write the x-coordinates of any inflection points. For each such x-coordinate, state why an inflection point occurs there. Confirm the results by observing the graph of f.

   (d) Use calculus to confirm these results by finding $f''(x)$, drawing its sign chart, and analyzing the results.

9. **Graph of f'' versus f.** Create a table and graphs for $f(x) = -x^3 - 6x^2 + 180x$ of Exercise 2 and its second derivative $f''(x) = -6x - 12$, following the format of Spreadsheet 10-5. Use the x-values -10, -9, -8, ..., 10.

(a) Observing the graph of f'', state the interval(s) on which f is concave up. Confirm the results by observing the graph of f.

(b) Observing the graph of f'', state the interval(s) on which f is concave down. Confirm the results by observing the graph of f.

(c) Observing the graph of f'', write the x-coordinates of any inflection points. For each such x-coordinate, state why an inflection point occurs there. Confirm the results by observing the graph of f.

(d) Use calculus to confirm these results by finding $f''(x)$, drawing its sign chart, and analyzing the results.

10. **Graph of f'' versus f: Acceleration and deceleration.** Create a table and graphs for $f(x) = 3x^4 - 8x^3 + 20$ of Exercise 3 and its second derivative $f''(x) = 36x^2 - 48x$, following the format of Spreadsheet 10-5. Use the x-values -1, -0.5, 0, 0.5, 1, 1.5, 2, 2.5, 3.

(a) Observing the graph of f'', state the interval(s) on which f is accelerating. Confirm the results by observing the graph of f.

(b) Observing the graph of f'', state the interval(s) on which f is decelerating. Confirm the results by observing the graph of f.

(c) Observing the graph of f'', write the x-coordinates of any inflection points. For each such x-coordinate, state why an inflection point occurs there. Confirm the results by observing the graph of f.

(d) Use calculus to confirm these results by finding $f''(x)$, drawing its sign chart, and analyzing the results.

11. **$f'' = 0$ does not always imply an inflection point.** Create a table and graphs for $f(x) = x^4$ and its second derivative $f''(x) = 12x^2$, following the format of Spreadsheet 10-5. Use the x-values -4, -3, -2, -1, 0, 1, 2, 3, 4.

(a) Observing the graph of f'', state the interval(s) on which f is concave up. Confirm the results by observing the graph of f.

(b) Observing the graph of f'', state the interval(s) on which f is concave down. Confirm the results by observing the graph of f.

(c) Observing the graph of f'', state why f has no inflection points. Confirm the results by observing the graph of f.

(d) Use calculus to confirm these results by finding $f''(x)$, drawing its sign chart, and analyzing the results.

12. *f" versus f.* Create a table and graph for $f''(x) = x^2 - 4$ using the x-values -4, -3, -2, 1, 0, 1, 2, 3, 4.

   (a) Observing the graph of f'', state the interval(s) on which f is concave up.

   (b) Observing the graph of f'', state the interval(s) on which f is concave down.

   (c) Observing the graph of f'', write the x-coordinates of any inflection points. For each such x-coordinate, state why an inflection point occurs there.

13. *f" versus f.* <u>Pencil and Paper Exercise</u>. Draw the graph of $f''(x) = -x^2 + 36$.

   (a) Observing the graph of f'', state the interval(s) on which f is concave up.

   (b) Observing the graph of f'', state the interval(s) on which f is concave down.

   (c) Observing the graph of f'', write the x-coordinates of any inflection points. For each such x-coordinate, state why an inflection point occurs there.

14. *f" versus f.* <u>Pencil and Paper Exercise</u>. Draw the graph of $f''(x) = (x-4)^2$.

   (a) Observing the graph of f'', explain why f is <u>not</u> concave down.

   (b) Observing the graph of f'', state the interval(s) on which f is concave up.

   (c) Explain why an inflection point does <u>not</u> occur at x = 4.

15. **Summary.** *Deriving information from a graph of f".* <u>Pencil and Paper Exercise</u>. When observing a graph of f'', explain how one determines:

   (a) Where f is concave up.   (b) Where f is concave down.   (c) Where f is accelerating.
   (d) Where f is decelerating. (e) Where inflection points occur.

16. <u>Pencil and Paper Exercise</u>. Draw a section of a curve that is:
   (a) Both increasing and concave down. (b) Both increasing and concave up.
   (c) Both decreasing and concave down. (d) Both decreasing and concave up.

17. <u>Pencil and Paper Exercise</u>. Draw a section of a curve that is:
   (a) Both increasing and accelerating. (b) Both increasing and decelerating.
   (b) Both decreasing and decelerating. (d) Both decreasing and accelerating.

18. *Graph of f" versus f'.* Spreadsheet 10-6 gives tables and graphs of
$f(x) = -2x^3 + 27x^2 - 108x + 165$, $f'(x) = -6x^2 + 54x - 108$, and
$f''(x) = -12x + 54$.

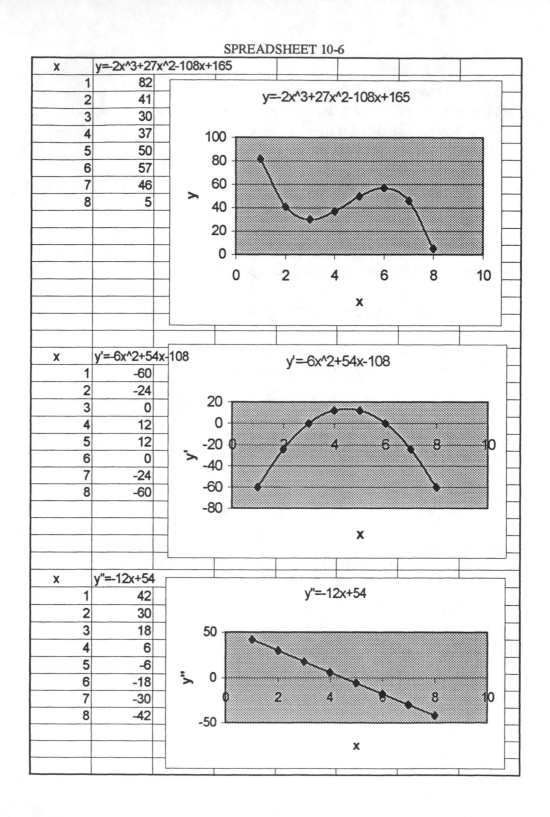

x	y=-2x^3+27x^2-108x+165
1	82
2	41
3	30
4	37
5	50
6	57
7	46
8	5

x	y'=-6x^2+54x-108
1	-60
2	-24
3	0
4	12
5	12
6	0
7	-24
8	-60

x	y"=-12x+54
1	42
2	30
3	18
4	6
5	-6
6	-18
7	-30
8	-42

Analogously, f'' *reveals about* f' *what* f' *reveals about* f. Observe the graph of f'' in Spreadsheet 10-6 and answer the following.

(a) State the interval(s) on which f' is increasing. Confirm the results by observing the graph of f'.

(b) State the interval(s) on which f' is decreasing. Confirm the results by observing the graph of f'.

(c) State the value of x at which f' has a local maximum. Explain why f is increasing most rapidly at this point.

19. *Graph of* f'' *versus* f'. Consider the second derivative $f''(x) = -6x + 18$.

(a) Use a spreadsheet to graph f'' using the x-values 0, 1, 2, 3, 4, 5, 6.

(b) Observing the graph of f'', state the interval(s) on which f' is increasing.

(c) Observing the graph of f'', state the interval(s) on which f' is decreasing.

(d) State the x-value at which f is increasing most rapidly.

20. *Graph of* f'' *versus* f'. Pencil and Paper Exercise. Consider the second derivative $f''(x) = 5x + 30$.

(a) Graph f''.

(b) Observing the graph of f'', state the interval(s) on which f' is increasing.

(c) Observing the graph of f'', state the interval(s) on which f' is decreasing.

(d) State the x-value at which f is decreasing most rapidly.

21. *Graph of* f'' *versus* f': *Profit function*. Pencil and Paper Exercise. Consider the profit function $P(x) = -0.01x^3 + 1.20x^2 - 21x + 50{,}000$, where $P(x)$ denotes the profit (in dollars) gained from selling x units.

(a) Find P'' and draw its graph.

(b) Observing the graph of P'', state the interval(s) on which P' is increasing.

(c) Observing the graph of P'', state the interval(s) on which P' is decreasing.

(d) State the x-value at which P is increasing most rapidly.

22. *Using* f' *and* f'' *to verify the graph of* f. Create a table and graph for $f(x) = x^{\frac{2}{3}}$ Use the x-values -1.10, -1.00, -0.90, . . . 1.10. Assuming the first x-value is in cell A2, type the formula =A2+0.10 in cell A3 and copy it down through the column until the last x-value is 1.10. Also, you might have to type the formula =(A2^2)^(1/3) instead of =A2^(2/3) to create the column of y-values.

(a) Pencil and Paper Exercise. Find f' and analyze its sign to determine where f increases, where it decreases, and where it has local maxima or minima.

(b) Pencil and Paper Exercise. Explain why f has a local minimum and why the graph comes to a sharp point (i.e., a cusp) at the local minimum.

(c) Pencil and Paper Exercise. Find f'' and analyze its sign to verify the concavity of f.

10-3 Optimization

Minimizing Average Cost per Unit.
Let $C(x)$ denote the total cost of producing x units of some item, where
$C(x) = x^2 - 60x + 1600$. As we learned in Section 3-2, *the average cost function* is given by

$$A(x) = \frac{C(x)}{x} \qquad\qquad (x > 0)$$

which in this case is

$$A(x) = x - 60 + \frac{1600}{x} \qquad\qquad (x > 0).$$

Note that this average cost equation was obtained by dividing each term of the cost equation by x. Thus, the average cost function gives the average cost per unit for various production levels, x. *Our goal is to determine the production level that minimizes the average cost per unit.*

By referring to Spreadsheet 10-7, the following instructions show how Excel's SOLVER is used to solve such an optimization problem.

SPREADSHEET 10-7

x	A(x)
1	1541

INSTRUCTIONS FOR SPREADSHEET 10-7

1. After typing the labels as indicated, type the number **1** in cell A2 and the formula **=A2-60+1600/A2** in cell B2. We're letting cell A2 contain x-values and cell B2 contain the corresponding y-values. We typed **1** in cell A2 to avoid division by 0 because of x in the denominator.

2. Use the mouse to *move the dark-bordered rectangle to the cell containing the formula for the function to be optimized*—in this case B2, which currently contains the value 1541.

3. Select *Tools* and then choose *Solver*, and a dialog box appears with the formula cell—in this case B2, identified in the *Set Target Cell* box as **B2**. This cell contains the formula the Solver tries to maximize or minimize. Click on *Min* to minimize. Then click on the white box under *By Changing Cells* and enter **A2** because cell A2 contains the x-values that are to be changed in order to minimize the average cost function formula in cell B2.

4. Click *Options* to make certain that a checkmark does <u>not</u> appear next to *Assume Linear Model*; our function is <u>not</u> linear. Click *Solve*, and SOLVER determines the optimal

values, if they exist. In this example, the *x*-value 40 appears in cell A2 and the *y*-value 20 appears in cell B2. Thus, the *local minimum point* is **(40, 20)** and this means that we should produce 40 units in order to achieve the minimum average cost per unit of $20. This is the optimal solution to our problem.

Spreadsheet 10-8 gives a table and graph of the average cost function

$$A(x) = x - 60 + \frac{1600}{x}$$ with $x > 0$. Graphs of this type of function are discussed in

Sections 3-1 and 3-2.

SPREADSHEET 10-8

x	A(x)
10	110
20	40
30	23.33333
40	20
50	22
60	26.66667
70	32.85714
80	40
90	47.77778
100	56
110	64.54545
120	73.33333
130	82.30769
140	91.42857
150	100.6667
160	110

Caution! One is tempted to question the need for SOLVER because the graph and table of *x*- and *y*-values in Spreadsheet 10-8 appear to have revealed the local minimum point without using SOLVER. However, this is dependent on the choice of *x*-values, and it is *only by chance* that the selected *x*-values in the table include the local minimum. There is no guarantee the *x*-values we select to create a table and graph will include the local maximum or minimum points. Thus, SOLVER obtains for us the exact local maximum or minimum point.

EXERCISES

1. Minimizing average cost per unit. . Let $C(x)$ denote the total cost of producing *x* units of some item, where $C(x) = x^2 - 80x + 3600$.

(a) Determine the equation of the *average cost function*.
(b) Use SOLVER to determine how many units should be produced in order to minimize the average cost per unit.
(c) Use a spreadsheet to create a table and graph for the average cost function. Use the *x*-

values 10, 20, 30, 40, 50, 60, 70, 80, 90, 100, 110, 120, 130 and confirm that SOLVER's answer is reasonable.

(d) Use f' and f'' to verify the local minimum.

2. Minimizing inventory cost. If inventory is ordered in batches of x units per order, the total annual inventory cost is given by $C(x) = 2x + \dfrac{2,000,000}{x}$, where $x > 0$.

(a) Use SOLVER to determine the order size that minimizes the total annual inventory cost.

(b) Return to Spreadsheet 3-5 in Section 3-2 and confirm that SOLVER's answer is reasonable.

(c) Use f' and f'' to verify the local minimum.

3. Minimizing inventory cost. If inventory is ordered in batches of x units per order, the total annual inventory cost is given by $C(x) = 8x + \dfrac{8,000,000}{x}$, where $x > 0$.

(a) Use SOLVER to determine the order size that minimizes the total annual inventory cost.

(b) Return to homework Exercise 5 in Section 3-2 and confirm that SOLVER's answer is reasonable.

(c) Use f' and f'' to verify the local minimum.

4. Maximizing profit. Let $P(x)$ denote the total profit gained from selling x units of some product for which $P(x) = -x^3 + 6x^2 + 36x - 10$, where $0 \le x \le 9$.

(a) Use SOLVER to determine how many units should be sold in order to maximize profit. Disregard the interval $0 \le x \le 9$.

(b) Use a spreadsheet to create a table and graph for the profit function, $P(x)$. Use the x-values 0, 1, 2, 3, 4, 5, 6, 7, 8, 9 and confirm that SOLVER's answer is reasonable.

(c) Use f' and f'' to verify the local maximum.

5. Let $f(x) = -x^3 + 6x^2 + 36x - 10$, where f has the same equation (without the restriction $0 \le x \le 9$) as the profit function of Exercise 4.

(a) Use SOLVER to minimize f.

(b) Use a spreadsheet to create a table and graph for the profit function, $P(x)$. Use the x-values -3, -2, -1, 0, 1, 2, 3, 4, 5, 6, 7, 8, 9 and confirm that SOLVER's answer is reasonable.

(c) Use f' and f'' to verify the local minimum.

6. Let $f(x) = x^3 - 3x + 2$.

(a) Use SOLVER to minimize f.

(b) Use SOLVER to maximize f.

(c) Use a spreadsheet to create a table and graph for f. Use the x-values -3, -2, -1, 0, 1, 2, 3 and confirm that SOLVER's answers are reasonable.

(d) Use f' and f'' to verify the local minimum and maximum.

CHAPTER ELEVEN

Definite Integrals

11-1 Rate-of-Change Step Functions

A *rate function* gives the *rate of change, y'*, of some quantity y, versus an independent variable. Typically, we must determine the *total change* in the quantity y on some interval.

One example of a rate function is a *velocity function*, which gives *velocity* (a *rate of change of distance with respect to time*) versus *time*. Because velocity is a rate of change of distance, y, with respect to time, a typical problem involves determining the total distance traveled during some time interval. Spreadsheet 11-1 gives a velocity function showing the velocity of a moving car versus time, t.

SPREADSHEET 11-1

t	v
0	30
1	30
2	30
3	30
4	30
4	60
5	60
6	60
7	60
8	60

Studying Spreadsheet 11-1, note that the car's velocity is 30 mph for the first 4 hours; at $t = 4$, the velocity instantaneously spikes up to 60 mph and remains at that level through the 8th hour. Thus, the graph of the velocity function resembles a step and is appropriately termed a *step function*.

Note: Observe the table of t- and v-values to note that $t = 4$ is listed twice: once with a velocity of 30 and then with a velocity of 60 to create the step at $t = 4$. It is useful to note this when creating graphs for some of the other rate-of-change step functions in the exercises. Although this technique associates two *y*-values for a given *x* at each step and therefore does <u>not</u> result in a function, we tolerate this in the interest of creating a useful graph.

Our problem is to determine the *total distance*, *y*, traveled by the car during the time interval from $t = 2$ to $t = 6$. The **total distance** is given by the **area under the graph of the velocity function**, and this area is illustrated by the two white rectangles under the graph of the velocity function in Spreadsheet 11-1. This is explained by noting that during the 2-hour time interval from $t = 2$ to $t = 4$, the car travels at a constant velocity of 30 mph and therefore travels a total distance of $(2)(30) = 60$ miles. Note that this result gives the area of the white rectangle on the interval from $t = 2$ to $t = 4$. During the second 2-hour time interval from $t = 4$ to $t = 6$, the car travels at a constant velocity of 60 mph and therefore travels a total distance of $(2)(60) = 120$ miles. Note that this result gives the area of the white rectangle on the interval from $t = 4$ to $t = 6$. Thus, the *total distance* traveled by the car on the interval from $t = 2$ to $t = 6$ is given by the sum of the areas of the two rectangles, or $60 + 120 = 180$ miles.

EXERCISES

1. *Velocity function.* Spreadsheet 11-2 gives a velocity function (in mph) for a moving car.

SPREADSHEET 11-2

t	v
0	20
1	20
2	20
2	50
3	50
4	50
5	50
6	50
6	80
7	80
8	80

(a) <u>Pencil and Paper Exercise</u>. State the car's velocity on the interval $0 < t < 2$.
(b) <u>Pencil and Paper Exercise</u>. State the car's velocity on the interval $2 < t < 6$.
(c) <u>Pencil and Paper Exercise</u>. State the car's velocity on the interval $6 < t < 8$.
(d) <u>Pencil and Paper Exercise</u>. Determine the total distance traveled by the car on the time interval from $t = 1$ to $t = 4$ and show the corresponding area under the graph.
(e) <u>Pencil and Paper Exercise</u>. Determine the total distance traveled by the car on the time interval from $t = 1$ to $t = 7$ and show the corresponding area under the graph.

2. *Velocity function.* A car travels at a constant velocity of 20 mph for the first 3 hours; at $t = 3$, the velocity instantaneously spikes up to 40 mph, and it remains at that level for

204

the next 2 hours; at $t = 5$, the velocity instantaneously spikes up to 80 mph, and it remains at that level for the next 3 hours.

(a) Use a spreadsheet to create a table and graph for this velocity function.

(b) Pencil and Paper Exercise. Determine the total distance traveled by the car on the time interval from $t = 1$ to $t = 4$ and show the corresponding area under the graph.

(c) Pencil and Paper Exercise. Determine the total distance traveled by the car on the time interval from $t = 1$ to $t = 7$ and show the corresponding area under the graph.

(d) Pencil and Paper Exercise. Determine the total distance traveled by the car on the time interval from $t = 1$ to $t = 8$ and show the corresponding area under the graph.

3. *Marginal cost function.* A marginal cost function, $C'(x)$, gives the rate of change of cost (\$) with respect to x, the number of units produced. Therefore, it is another example of a rate function. Spreadsheet 11-3 gives a marginal cost function for some company, where x denotes the number of units produced in thousands. In other words, $x = 3$ means that 3 thousand units are produced.

SPREADSHEET 11-3

x	C'(x)
0	10
1	10
2	10
3	10
3	8
4	8
5	8
6	8
6	4
7	4
8	4

(a) Pencil and Paper Exercise. State the marginal cost for the first 3 thousand units produced (i.e., on the interval $0 < x < 3$).

(b) Pencil and Paper Exercise. State the marginal cost for the next 3 thousand units produced (i.e., on the interval $3 < x < 6$).

(c) Pencil and Paper Exercise. State the marginal cost on the interval $6 < x < 8$.

(d) Pencil and Paper Exercise. Determine the *total cost* of producing the additional units when the production level increases from $x = 1$ to $x = 4$ and show the corresponding area under the graph. This also means the *total change (increase) in cost* when production increases from $x = 1$ to $x = 4$.

(e) Pencil and Paper Exercise. Determine the *total cost* of producing the additional units when the production level increases from $x = 4$ to $x = 7$ and show the corresponding area under the graph. This also means the *total change (increase) in cost* when production increases from $x = 4$ to $x = 7$.

(f) Pencil and Paper Exercise. Determine the *total cost* of producing the additional units when the production level increases from $x = 2$ to $x = 6$ and show the corresponding area under the graph. This also means the *total change (increase) in cost* when production increases from $x = 2$ to $x = 6$.

Summary

A *rate function* gives the *rate of change, y'* , of some quantity y, versus an independent variable, say x. If y' is positive, the *total change in* y on the interval from $x = a$ to $x = b$ is given by the *area under the graph of y'* over the interval $a < x < b$.

4. *Marginal cost function.* As we learned in Exercise 3, a marginal cost function, $C'(x)$, gives the rate of change of cost with respect to x, the number of units produced.

A firm's marginal cost is $20 per unit for the first 4 thousand units produced, $10 per unit for the next 5 thousand units produced, and $5 per unit for the next 2 thousand units produced.

  (a) Use a spreadsheet to create a table and graph for the marginal cost function.
  (b) Pencil and Paper Exercise. Determine the *total change (increase) in cost* when production increases from $x = 1$ to $x = 3$ and show the corresponding area under the graph.
  (c) Pencil and Paper Exercise. Determine the *total change (increase) in cost* when production increases from $x = 4$ to $x = 7$ and show the corresponding area under the graph.
  (d) Pencil and Paper Exercise. Determine the *total change (increase) in cost* when production increases from 2 thousand to 6 thousand units and show the corresponding area under the graph.
  (e) Pencil and Paper Exercise. Determine the *total change (increase) in cost* when production increases from 3 thousand to 8 thousand units and show the corresponding area under the graph.

Summary. For a positive marginal cost function, $C'(x)$, when the production level increases from $x = a$ to $x = b$, the cost of making the additional units is given by the area under the graph of $C'(x)$ over the interval $a < x < b$.

5. *Marginal profit function.* A marginal profit function, $P'(x)$, gives the rate of change of profit ($) with respect to x, the number of units produced and sold. Spreadsheet 11-4 gives a marginal profit function for some company, where x denotes the number of units produced and sold in thousands. In other words, $x = 3$ means that 3 thousand units are produced and sold.

  (a) Pencil and Paper Exercise. Studying the marginal profit function of Spreadsheet 11-4, state the marginal profit for the first 3 thousand units produced and sold (i.e., on the interval $0 < x < 3$).
  (b) Pencil and Paper Exercise. State the marginal profit for the next 4 thousand units produced and sold (i.e., on the interval $3 < x < 7$).
  (c) Pencil and Paper Exercise. State the marginal profit on the interval $7 < x < 9$.
  (d) Pencil and Paper Exercise. Determine the *total profit* gained from producing and selling the additional units when the sales level increases from $x = 1$ to $x = 4$ and show the corresponding area under the graph. This also means the *total change (increase) in profit* when sales increase from $x = 1$ to $x = 4$.

(e) <u>Pencil and Paper Exercise</u>. Determine the *total profit* gained from producing and selling the additional units when the sales level increases from $x = 4$ to $x = 7$ and show the corresponding area under the graph. This also means the *total change (increase) in profit* when sales increase from $x = 4$ to $x = 7$.

(f) <u>Pencil and Paper Exercise</u>. Determine the *total profit* gained from producing and selling the additional units when the sales level increases from $x = 2$ to $x = 6$ and show the corresponding area under the graph. This also means the *total change (increase) in profit* when sales increase from $x = 2$ to $x = 6$.

SPREADSHEET 11-4

x	P'(x)
0	5
1	5
2	5
3	5
3	8
4	8
5	8
6	8
7	8
7	12
8	12
9	12

6. *Marginal profit function.* A firm's marginal profit function is such that the first 5 thousand units produced and sold yield a profit $4 per unit; the next 4 thousand units yield a profit of $7 per unit; the next 2 thousand units yield a profit of $9 per unit.

(a) Use a spreadsheet to create a table and graph for the marginal profit function.

(b) <u>Pencil and Paper Exercise</u>. Determine the *total profit* gained from producing and selling the additional units when the sales level increases from $x = 2$ to $x = 4$ and show the corresponding area under the graph. This also means the *total change (increase) in profit* when sales increase from $x = 2$ to $x = 4$.

(c) <u>Pencil and Paper Exercise</u>. Determine the *total profit* gained from producing and selling the additional units when the sales level increases from $x = 4$ to $x = 7$ and show the corresponding area under the graph. This also means the *total change (increase) in profit* when sales increase from $x = 5$ to $x = 7$.

(d) <u>Pencil and Paper Exercise</u>. Determine the *total profit* gained from producing and selling the additional units when the sales level increases from $x = 2$ to $x = 6$ and show the corresponding area under the graph. This also means the *total change (increase) in profit* when sales increase from $x = 4$ to $x = 8$.

7. *Marginal tax rates.* A marginal tax rate gives the amount of income tax due on each additional dollar of taxable income. Spreadsheet 11-5 gives the marginal tax rates for taxable income, x, up to 100 thousand dollars, where x denotes taxable income in thousands of dollars. In other words, $x = 20$ means taxable income of 20 thousand dollars.

SPREADSHEET 11-5

x	Marg tax rte
0	0.1
10	0.1
20	0.1
20	0.15
30	0.15
40	0.15
50	0.15
60	0.15
70	0.15
70	0.2
80	0.2
90	0.2
100	0.2

Marginal tax rates

(chart of marginal tax rate vs. x, with y-axis from 0 to 0.2 and x-axis from 0 to 100)

(a) <u>Pencil and Paper Exercise</u>. State the marginal tax rate for the first 20 thousand dollars of taxable income (i.e., on the interval $0 < x < 20$).

(b) <u>Pencil and Paper Exercise</u>. State the marginal tax rate for the next 50 thousand dollars of taxable income (i.e., on the interval $20 < x < 70$).

(c) <u>Pencil and Paper Exercise</u>. State the marginal cost on the interval $70 < x < 100$.

(d) <u>Pencil and Paper Exercise</u>. Determine the *total tax* on taxable income of 15 thousand dollars and show the corresponding area under the graph.

(e) <u>Pencil and Paper Exercise</u>. Determine the *total tax* on taxable income of 60 thousand dollars and show the corresponding area under the graph.

(f) <u>Pencil and Paper Exercise</u>. Determine the *total tax* on taxable income of 90 thousand dollars and show the corresponding area under the graph.

(g) <u>Pencil and Paper Exercise</u>. Determine the *total change (increase) in tax* when taxable income increases from $x = 20$ to $x = 30$ and show the corresponding area under the graph.

(h) <u>Pencil and Paper Exercise</u>. Determine the *total change (increase) in tax* when taxable income increases from 40 to 60 thousand dollars and show the corresponding area under the graph.

(i) <u>Pencil and Paper Exercise</u>. Determine the *total change (increase) in tax* when taxable income increases from 60 to 80 thousand dollars and show the corresponding area under the graph.

8. *Marginal tax rates.* A city's marginal tax rates are as follows. The first 20 thousand dollars of taxable income are taxed at 12%; the next 50 thousand dollars at 16%; and the next 30 thousand up to 100 thousand dollars at 25%.

(a) Use a spreadsheet to create a table and graph of the marginal tax rates.

(b) <u>Pencil and Paper Exercise</u>. Determine the *total tax* on taxable income of 80 thousand dollars and show the corresponding area under the graph.

(c) <u>Pencil and Paper Exercise</u>. Determine the *total change (increase) in tax* when taxable income increases from 40 to 80 thousand dollars and show the corresponding area under the graph.

11-2 Riemann Sums

In Section 11-1, we determined areas under graphs of rate functions that are step functions. Thus, our areas have been those of rectangles or sums of rectangles. In this section, we use rectangles to *approximate* area under a curve $y = f(x)$ on some interval from $x = a$ to $x = b$. Spreadsheet 11-6 illustrates such an approximation for $f(x) = x^2$ on the interval from $x = 0$ to $x = 1$.

SPREADSHEET 11-6

x	y = x^2					
0	0					
0.25	0.0625					
0.5	0.25					
0.75	0.5625					
1	1					

x	f(x)	dx	f(x) dx			
0	0					
0.25	0.0625	0.25	0.015625			
0.5	0.25	0.25	0.0625			
0.75	0.5625	0.25	0.140625			
1	1					
			0.21875	←—Lower approximation:		
				Sum of 3 rectangles		

Observing the graph in the upper portion of Spreadsheet 11-6, note that three white rectangles are used to approximate the area under $f(x) = x^2$ on the interval from $x = 0$ to $x = 1$. The *width* of each rectangle is 0.25 (denoted by dx), and the *height* is given by the y-coordinate of the point where upper left-hand corner of the rectangle touches the graph. The individual areas of the rectangles are computed in the bottom portion of the spreadsheet and are given in the **f(x) dx** column. The sum of these areas is **0.21875** and is given in boldface type. Because the sum of the rectangles gives an area that is *smaller* than the actual area, it is called a *lower approximation*.

209

Spreadsheet 11-7 gives an approximation that is *larger* than the actual area under $f(x) = x^2$ on the interval from $x = 0$ to $x = 1$ and is therefore called an *upper approximation*.

SPREADSHEET 11-7

x	y = x^2								
0	0								
0.25	0.0625								
0.5	0.25								
0.75	0.5625								
1	1								

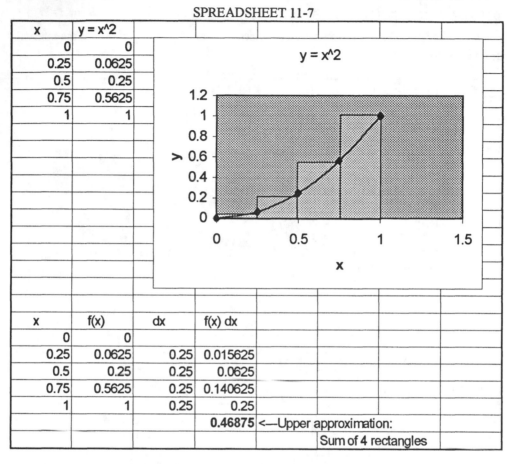

x	f(x)	dx	f(x) dx					
0	0							
0.25	0.0625	0.25	0.015625					
0.5	0.25	0.25	0.0625					
0.75	0.5625	0.25	0.140625					
1	1	0.25	0.25					
			0.46875	←—Upper approximation:				
				Sum of 4 rectangles				

As in Spreadsheet 11-6, the individual areas of the rectangles are computed in the bottom portion of the spreadsheet and are given in the **f(x) dx** column. The sum of these areas is **0.46875** and is given in boldface type.

Combining the results of Spreadsheets 11-6 and 11-7 yields

Lower approximation < actual area < upper approximation

0.21875 < actual area < 0.46875.

More accurate approximations are obtained by using more rectangles of thinner width as illustrated in Exercise 1. Such sums are called Riemann sums. As the width of each rectangle gets smaller and smaller so that the number of rectangles increases without bound, the limit of a Riemann sum gives the actual area. Of course, the *actual area* is given by $\int_0^1 x^2\,dx = \dfrac{1}{3}$.

210

EXERCISES

1. *Riemann sum.* Spreadsheet 11-8 illustrates the computation of both a lower and an upper approximation to the area under the graph of $f(x) = x^2$ on the interval from $x = 0$ to $x = 1$ using thinner rectangles with $dx = 0.125$. Of course, these approximations are more accurate than those determined in Spreadsheets 11-6 and 11-7.

SPREADSHEET 11-8

LOWER APPROXIMATION				UPPER APPROXIMATION			
x	f(x) = x^2	dx	f(x) dx	x	f(x) = x^2	dx	f(x) dx
0	0	0.125		0	0	0.125	
0.125	0.015625	0.125	0.001953	0.125	0.015625	0.125	0.001953
0.25	0.0625	0.125	0.007813	0.25	0.0625	0.125	0.007813
0.375	0.140625	0.125	0.017578	0.375	0.140625	0.125	0.017578
0.5	0.25	0.125	0.03125	0.5	0.25	0.125	0.03125
0.625	0.390625	0.125	0.048828	0.625	0.390625	0.125	0.048828
0.75	0.5625	0.125	0.070313	0.75	0.5625	0.125	0.070313
0.875	0.765625	0.125	0.095703	0.875	0.765625	0.125	0.095703
1	1	0.125		1	1	0.125	0.125
Lower approximation————>			0.273438	Lower approximation————>			0.398438

(a) <u>Pencil and Paper Exercise</u>. Observing the **f(x) dx** column in the LOWER APPROXIMATION section of Spreadsheet 11-8, state the x-coordinate of the left-hand side of the first approximating rectangle and the corresponding y-coordinate of the point where the rectangle touches the curve.

(b) <u>Pencil and Paper Exercise</u>. Observing the **f(x) dx** column in the LOWER APPROXIMATION section of Spreadsheet 11-8, state the x-coordinate of the left-hand side of the last approximating rectangle and the corresponding y-coordinate of the point where the rectangle touches the curve.

(c) <u>Pencil and Paper Exercise</u>. Observing the **f(x) dx** column in the UPPER APPROXIMATION section of Spreadsheet 11-8, state the x-coordinate of the left-hand side of the first approximating rectangle and the corresponding y-coordinate of the point where the rectangle touches the curve.

(d) <u>Pencil and Paper Exercise</u>. Observing the **f(x) dx** column in the UPPER APPROXIMATION section of Spreadsheet 11-8, state the x-coordinate of the left-hand side of the last approximating rectangle and the corresponding y-coordinate of the point where the rectangle touches the curve.

(e) <u>Pencil and Paper Exercise</u>. Verify that the interval between the lower and upper approximations of Spreadsheet 11-8 contains the actual area and state why it is a better (more accurate) approximation than that determined in Spreadsheets 11-6 and 11-7.

2. *Riemann sum.* Use a spreadsheet to create a table of x- and y-values and graph for $f(x) = x^3$ for the x-values -1, -0.75, -0.50, -0.25, 0, 0.25, 0.50, 0.75, 1.

(a) Use a spreadsheet to determine both lower and upper approximations to the area under $f(x) = x^3$ on the interval from $x = 0$ to $x = 1$ with $dx = 0.25$ as the width of each approximating rectangle.

(b) Use a spreadsheet to determine both lower and upper approximations to the area under $f(x) = x^3$ on the interval from $x = 0$ to $x = 1$ with $dx = 0.125$ as the width of each approximating rectangle.

(c) Pencil and Paper Exercise. Use the definite integral to determine the actual area under the curve.

(d) Pencil and Paper Exercise. Verify that the interval between the lower and upper approximations of parts (a) and (b) contains the actual area and state why the approximation of part (b) is a better (more accurate) approximation than that of part (a).

3. *Riemann sum: Area below the x-axis.* Use a spreadsheet to determine both lower and upper approximations to the area between the graph of $f(x) = x^3$ and the x-axis on the interval from $x = -1$ to $x = 0$. Refer to the graph created in Exercise 2, if needed.

(a) Use $dx = 0.25$ as the width of each approximating rectangle.

(b) Use $dx = 0.125$ as the width of each approximating rectangle.

(c) Pencil and Paper Exercise. Explain why the approximations are negative.

(d) Pencil and Paper Exercise. Use the definite integral to determine the actual area under the curve.

4. *Riemann sum: Velocity function.* The velocity (in feet per second) of a moving particle after t seconds is given by $v(t) = 4t^3$.

(a) Create a table and graph for this velocity function using the t-values 0, 0.5, 1, 1.5, 2, 2.5.

(b) Use a spreadsheet to determine both lower and upper approximations for the *total distance* traveled by the particle on the time interval from $t = 1$ to $t = 2$ with $dt = 0.25$.

(c) Use a spreadsheet to determine both lower and upper approximations for the *total distance* traveled by the particle on the time interval from $t = 1$ to $t = 2$ with $dt = 0.125$.

(d) Pencil and Paper Exercise. Use the definite integral to determine the actual area under the curve that gives the total distance traveled by the particle from $t = 1$ to $t = 2$.

5. *Riemann sum: Marginal cost function.* Consider the marginal cost function

$$C'(x) = \frac{500}{x}$$, where x denotes the number of units produced.

(a) Create a table and graph for this marginal cost function using the x-values 0.1, 0.5, 1, 1.5, 2, 2.5, 3, 4, 5, 6, 7, 8, 9, 10.

(b) Use a spreadsheet to determine both lower and upper approximations for the *total change (increase) in cost* when production increases from $x = 1$ to $x = 5$ with $dx = 0.25$.

(c) Pencil and Paper Exercise. Use the definite integral to determine the *actual area* under the curve that gives the *total increase in cost* approximated in part (b).

CHAPTER TWELVE

Optimization: Functions of Two Variables

12-1 Optimization

In this section, we use **Solver** to optimize functions, $z = f(x, y)$. As an example, we consider the function

$$f(x, y) = -x^2 + 8x - 2y^2 + 6y + 2xy + 50$$

where z is the sales revenue derived from selling x units of product A and y units of product B. We must determine how many units of each product should be sold in order to *maximize sales revenue*.

The following instructions, which refer to Spreadsheet 12-1, show how Excel's **Solver** is used to solve such an optimization problem.

SPREADSHEET 12-1

z	50				
x	0				
y	0				

INSTRUCTIONS FOR SPREADSHEET 12-1

1. After typing in labels as shown, type the number 0 in cells B2 and B3 and the formula =-1*B2^2+8*B2-2*B3^2+6*B3+2*B2*B3+50 in cell B1. Note that cells B2 and B3 contain x- and y-values, respectively, and typing **0** in these cells sets their initial values to 0. Also, the **-1** multiplier of **B2^2** in the formula ensures against Excel erroneously squaring the negative.

2. Use the mouse to *move the dark-bordered rectangle to the cell containing the formula for the function to be maximized*—in this case B1, which currently contains the value 50.

3. Select *Tools*, and then choose *Solver*, and a dialog box appears with the formula cell—in this case B1—identified in the *Set Target Cell* box as **B1**. This cell contains the formula that **Solver** tries to maximize or minimize. Click on *Max* to maximize and then click on the white box under *By Changing Cells* and enter **B2:B3** because cells B2 and B3 contain the independent variables, *x* and *y*, that are to be changed in order to maximize the formula for *z* in cell B1.

4. Click *Options* to make certain that a checkmark does not appear next to *Assume Linear Model*; our function is not linear. Click *Solve*, and **Solver** determines the optimal values, if they exist. In this example, the *x*-value 11 appears in cell B2, the *y*-value 7 appears in cell B3, and the *z*-value 115 appears in cell B1. Thus, 11 units of product A and 7 units of product B should be sold in order to achieve maximum sales revenue of $115. This is the optimal solution to our problem.

EXERCISES

For each of the following:

(a) Use SOLVER to optimize the given problem as indicated.
(b) <u>Pencil and Paper Exercise</u>. Use partial derivatives to verify SOLVER's result.

1. Minimize $f(x,y) = x^2 + 2y^2 - 8x - 20y + 18$.

2. Maximize $f(x,y) = 40x + 160y - 2x^2 - 4y^2 + 1000$.

3. Minimize $f(x,y) = 4x^2 + 2y^2 + 3xy - 70x - 55y + 1000$.

4. Maximize $f(x,y) = 200 - 2x^2 - 6y^2 + 2xy + 32x + 28y$.

5. Minimize $f(x,y) = 80 + 3xy + 42x - 16y - 3x^2 - 2y^2$.

6. *Maximizing profit.* The profit of a company is given by

$$P(x,y) = 1,000,000 + 1600x + 2000y - 4x^2 - 2y^2$$

where *x* is the unit labor cost and *y* is the unit raw material cost.
(a) Find the unit labor cost and unit raw material cost that maximize profit.
(b) Find the maximum profit.

7. *Minimizing cost.* The manager of a frogurt stand has determined that the cost of producing x gallons of strawberry frogurt and y gallons of blueberry frogurt is given by

$$C(x,y) = 2x^2 + 3y^2 + 2xy - 800x - 1400y + 185,000.$$

(a) How many gallons of each flavor should be produced in order to minimize cost?
(b) State the minimum cost.

8. *Maximizing production output.* The weekly output of a firm is given by

$$Z(x,y) = 1000x + 1600y + 2xy - 5x^2 - 2y^2$$

where x is the number of hours of labor and y is the number of units of raw material used weekly.
(a) How many hours of labor and how many units of raw material should be used weekly in order to maximize output?
(b) State the maximum output.

9. *Maximizing profit.* A firm produces two products, which are used in the automobile industry. Each thousand units of product 1 sells for $200, and each thousand units of product 2 sells for $295. If x thousand units of product 1 are produced and y thousand units of product 2 are produced, the total production cost is given by

$$C(x,y) = 5x^2 + 10y^2 + 5xy - 10x + 15y + 10.$$

(a) Pencil and Paper Exercise. Determine the equation for total sales revenue, $R(x,y)$.
(b) Pencil and Paper Exercise. Determine the equation for total profit, $P(x,y)$.
(c) How many units of each product should be produced in order to maximize total profit?
(d) State the maximum profit.

12-2 Method of Least Squares

In Section 1-4, we used Excel to insert a trendline to a set of data points. As we noted then, Excel's *Trendline* inserts the *best-fitting* line to a set of data points using the *Method of Least Squares*. In this section, we will use Excel's *Solver* in order to understand the optimization concepts behind the least-squares method.

SPREADSHEET 12-2

x	y		Residuals	ResidSQ		
x	y	Y=a + bx	y - Y	(y - Y)^2		
0	5	a + b(0)	5 - a	(5 - a)^2		
1	3	a + b(1)	3 - (a + b)	(3-a-b)^2		
2	6	a + b(2)	6 - (a+2b)	(6-a-2b)^2		
3	14	a + b(3)	14-(a+3b)	(14-a-3b)^2		

The top portion of Spreadsheet 12-2 gives a set of data points (x, y) and a plot of their graph, along with Excel's fitted trendline. Although we choose to fit a linear trendline, the least- squares method can be used to fit nonlinear trendlines as well. Spreadsheet 12-2 contains the same data points as those given in Spreadsheet 1-4, so we already know the equation of the fitted trendline because it is included in the graph of Spreadsheet 1-4.

However, in this section, we assume that we do not know the equation of the trendline and, instead, use the linear form $Y = a + bx$ to denote its equation. Thus, our objective is to determine the *y-intercept*, *a*, and *slope*, *b*, of the straight line that *best fits* the data points.

As stated in Section 1-4, the *goodness of fit* of a line fit to a set of points is quantified by considering, for each data point, the *vertical distance between the data point and the fitted line*. Such vertical distances are called *residuals* and are included as *vertical lines* drawn between the data points and the fitted line in the graph of Spreadsheet 12-2. The residuals indicate the extent to which the fitted line does not fit the data points. An overall measure of the extent to which a fitted line does not fit the data points is given by the *sum of the squares of the residuals*. Thus, the line that *minimizes the sum of the squares of the residuals* is the *best-fitting* line to the set of data points and is appropriately called the *least-squares line*.

216

The residuals are written in terms of the y-intercept, a, and slope, b, in the bottom portion of Spreadsheet 12-2. The column labeled y - Y (y denotes the data points; Y the fitted line) with **Residuals** appearing above contains the residuals; the column labeled **ResidSQ** contains the squares of the residuals. Because our goal is to minimize the sum of the squares of the residuals, we sum the entries of the ResidSQ column to obtain

$$S = (5-a)^2 + (3-a-b)^2 + (6-a-2b)^2 + (14-a-3b)^2$$

where our objective is to determine the values of a and b that **minimize** S.

Spreadsheet 12-3 and the following instructions illustrate how to use Excel's **Solver** to solve such an optimization problem.

SPREADSHEET 12-3

S	266			Method of Least Squares	
a	0			Same data points	
b	0			as Spreadsheet 1-	4

INSTRUCTIONS FOR SPREADSHEET 12-3

1. After typing in labels as shown, type the number 0 in cells B2 and B3 and the formula =(5 - B2)^2+(3 - B2 - B3)^2+(6 - B2 - 2*B3)^2+(14 - B2 - 3*B3)^2 in cell B1. This enters the function to be minimized. Note that cells B2 and B3 contain the a- and b-values, respectively, and typing 0 in these cells sets their initial values to 0.

2. Use the mouse to *move the dark-bordered rectangle to the cell containing the formula for the function to be minimized*—in this case B1, which currently contains the value 266.

3. Select *Tools*, and then choose *Solver*, and a dialog box appears with the formula cell— in this case B1—identified in the *Set Target Cell* box as **B1**. This cell contains the formula that **Solver** tries to maximize or minimize. Click on *Min* to minimize and then click on the white box under *By Changing Cells* and enter **B2:B3** because cells B2 and B3 contain the independent variables, a and b, that are to be changed in order to minimize the formula for S in cell B1.

4. Click *Options* to make certain that a checkmark does <u>not</u> appear next to *Assume Linear Model*; our function is <u>not</u> linear. Click *Solve*, and Solver determines the optimal values, if they exist. In this example, 2.5 appears in cell B2 as the value for a; 3 appears in cell B3 as the value for b; 25 appears in cell B1 as the minimum value of S. Comparing these results with those of Spreadsheet 1-4 confirms that they are the same. Thus, $y = 2.5 + 3x$ is the equation of the least-squares line.

EXERCISES

For each of the following sets of data:
  (a) Write the equation that gives the sum of the squares, S, of the residuals in terms of a and b, where $y = a + bx$ denotes the equation of the least-squares line.
  (b) Use **Solver** to determine the values of a and b that minimize S.
  (c) Write the equation of the least-squares line.
  (d) State the minimum value of S.

1.

x	y
2	10
5	15
8	25
9	30

2.

x	y
3	9
1	5
2	7
5	14

3.

x	y
2	8
4	14
8	30
9	40

4.

x	y
2	8
3	10
2	7
5	15

5.

x	y
3	7
2	5
4	10
3	6

6.

x	y
4	8
2	5
8	14
6	9

12-3 Constrained Optimization: Lagrange Multipliers

Sometimes we must optimize a function $z = f(x, y)$ where x and y are constrained. Consider a factory that burns two types of fuel: F108 and F109. The number of tons of pollutants emitted by the factory in a year is given by

$$f(x,y) = x^2 + 2y^2 - xy - 279{,}990$$

where x is the amount (in thousands of gallons) of F108 fuel used annually and y is the amount (in thousands of gallons) of F109 fuel used annually. The factory uses a combined amount of 800 thousand gallons of fuel annually. We seek to *determine how many thousands of gallons of each type of fuel should be burned annually in order to minimize the amount of pollutant exhausted.*

The factory uses a combined amount of 800 thousand gallons of fuel annually, so

$$x + y = 800 \,.$$

Mathematically, our problem is to

Minimize $f(x,y) = x^2 + 2y^2 - xy - 279{,}990$
subject to the constraint $g(x,y) = x + y = 800$.

Such a problem is solved by the method of *Lagrange multipliers.*

In general, the method of Lagrange multipliers is used to solve the following type of problem.

Maximize (or minimize) $z = f(x,y)$
subject to $g(x,y) = c$ where c is a constant.

The method of Lagrange multipliers involves defining and optimizing a new function

$$F(x,y,\lambda) = f(x,y) + \lambda(c - g(x,y))$$

where λ is called the **Lagrange multiplier** and the function F is called the Lagrangian function. Studying the Lagrangian function, F, note that because $g(x,y) = c$, then $c - g(x,y) = 0$, and the value of F will equal that of the original function, f.

It is proven in more advanced texts that finding the values of x, y, and λ that maximize (or minimize) F also maximize (or minimize) $f(x,y)$ subject to the constraint $g(x,y) = c$.

Here, we show how to use **Solver** for such constrained optimization problems. Spreadsheet 12-4 illustrates the entries for the above problem.

SPREADSHEET 12-4

MINIMIZATION					
Obj z	-279990				
x	0				
y	0				
Constraint	0	800			

INSTRUCTIONS

Our spreadsheet format is a simplified version of that used for linear programming.

1. Type Labels as Indicated

2. Enter Formulas and Initial Values
2.1 Because we're using cells B5 and B6 to contain the x- and y-values, respectively, our formula is written in terms of cells B5 and B6. Enter the formula **=B5^2+2*B6^2-B5*B6-279990** for the *objective function* in cell B3. The constant -279990 appears in cell B3 because cells B5 and B6 have no values at this point. Type **0** for the initial values of x and y in cells B5 and B6. As indicated, these cells contain the values of the decision variables, x and y, respectively.

2.2 Enter the formula **=B5+B6** for the *left-hand side of the constraint* in cell B8 and type the *right-hand side constant*, **800**, in cell C8. A zero will appear in cell B8.

3. Use Solver

(Enter the objective function.)
3.1 Use the mouse to *move the dark-bordered rectangle to the cell containing the formula for the objective function*—in this case cell B3, which currently contains the value 0.

3.2 Select *Tools*, then choose *Solver*, and a dialog box appears with the *objective function cell*, B5, identified in the *Set Target Cell* box as **B3**. This cell contains the formula that Solver tries to maximize or minimize. Click on *Min* to minimize, click on the white box under *By Changing Cells*, and either select the cells (using the mouse) containing the values of x and y, cells B5 and B6, or type in **B5:B6**.

(Enter the constraint.)
3.3 Click *Add* to the right of the white box appearing under *Subject to the Constraints*, and a dialog box appears. Either select the cell (using the mouse)—in this case B8—

220

containing the formula for the constraint and click inside the white box under *Cell Reference*, or type **B8** inside the white box. Choose the appropriate inequality symbol, in this case =, and again either select the cell (using the mouse) containing the right-hand-side constant and click inside the white box located to the right of the inequality symbol, or type **C8** inside the white box. Click *Add* to include this constraint.

3.4 Because this includes all the constraints, click *Cancel*, and the original **Solver** dialog box appears with a listing of the cells corresponding to the constraints.

(Obtain the optimal solution.)
3.5 Click *Options* to make certain that a checkmark does <u>not</u> appear next to *Assume Linear Model*; our function is <u>not</u> linear. Click **OK** to return to the **Solver** dialog box.

3.6 Click *Solve*, and **Solver** determines optimal values (if they exist) for the decision variables. Note that **Solver** changes the values in your spreadsheet by replacing them with the optimal solution as illustrated in Spreadsheet 12-5.

SPREADSHEET 12-5

MINIMIZATION							
Obj Z	10						
x	500						
y	300						
Constraint	800	800					

Studying Spreadsheet 12-5 reveals that the *optimal solution* is $x = 500$ and $y = 300$ to give a *minimum objective function value* of $Z = 10$.

3.7 Use the **Solver** Results dialog box to prepare a report that details the results of the optimal solution on the problem. Double-click *Sensitivity* in the *Reports* box and click **OK**. Finally, click on *Sensitivity report* at the bottom of your screen to obtain the report. The report for our illustrative problem is given on the following page.

Observe, on the next page, that the Sensitivity Report gives the optimal values of x and y along with the Lagrange multiplier, λ. Thus, the optimal solution including the Lagrange multiplier is
$$x = 500 \quad y = 300 \quad \lambda = 700.$$

Interpretation of the Lagrange Multiplier, λ
The Lagrange multiplier, $\lambda = 700$, gives the *rate of change of the objective function z with respect to the constraint value, c*. In other words, it gives the sensitivity of the optimal value, z, of the objective function to a change in c. For this example, each unit increase in c (i.e., each additional thousand gallons of fuel used) increases the optimal amount of pollutants emitted by approximately 700 tons.

SENSITIVITY REPORT

Changing Cells

Cell	Name	Final Value	Reduced Gradient
$B$5	x	500	0
$B$6	y	300	0

Constraints

Cell	Name	Final Value	Lagrange Multiplier
$B$8	Constrain	800	700.000473

EXERCISES

1 – 8. For each of the following, use **Solver** to optimize the given problem as indicated.

1.   Minimize $f(x,y) = x^2 - 4xy + y^2 + 200$
         subject to the constraint $2x + y = 26$.

2.   Minimize $f(x,y) = x^2 + 6xy + y^2$
         subject to the constraint $x + y = 10$.
         Also, interpret the Lagrange multiplier, λ .

3.   Minimize $f(x,y) = x^2 + 6xy + 2y^2$
         subject to the constraint $4x + y = 18$.

4.   Maximize $f(x,y) = x^2 + 4xy + y$
         subject to the constraint $x + y = 12$.

5.   Maximize $f(x,y) = -y^2 + xy + x$
         subject to the constraint $2x + y = 19$.
         Also, interpret the Lagrange multiplier, λ .

6.   Minimize $f(x,y) = x^2 - 5xy + 2y^2$
         subject to the constraint $3x + y = 20$.

7. *Maximizing profit.* A farmer's profit per square foot of cropland is given by

$$P(x,y) = -x^2 - 5y^2 + 10xy + 4x + 2y - 1100$$

where x is the amount spent on labor per square foot and y is the amount spent on fertilizer per square foot of cropland. If the farmer spends a total of $31.90 per square foot of cropland for labor and fertilizer, how many dollars per square foot should be allocated to labor and to fertilizer in order to maximize the profit per square foot?

8. *Minimizing cost.* The total cost of producing x units of product A and y units of product B is given by

$$C(x,y) = x^2 + 4y^2 - 5xy + 2000.$$

If a combined total of 40 units is produced daily, how many units of each product should be produced daily in order to minimize the total cost?

9. *Interpret* λ. Interpret the Lagrange multiplier, λ, for the spreadsheet result of Exercise 7.

10. *Interpret* λ. Interpret the Lagrange multiplier, λ, for the spreadsheet result of Exercise 8.

ANSWERS TO SELECTED EXERCISES

CHAPTER 0

1. The completed Spreadsheet 0-5 is given below.

	A	B	C	D	E	F	G	H
1		Currency per US $				US $ equivalent		
2		E	L	%Ch		1/E	1/L	%Ch
3	peso	0.99	0.98	-1.01%		1.010101	1.020408	1.02%
4	real	1.7695	1.752	-0.99%		0.565131	0.570776	1.00%
5	krone	7.198	7.17	-0.39%		0.138927	0.13947	0.39%
6	guilder	2.1333	2.1587	1.19%		0.468757	0.463242	-1.18%

(b) $1 buys 1.7695 real or, equivalently, $1 = 1.7695 real.
   $1 buys 1.752 real or, equivalently, $1 = 1.752 real.
(d) During the period between the early and late points in time, the dollar has depreciated by 1.01% against the peso.
(e) During the period between the early and late points in time, the dollar has appreciated by 1.19% against the guilder.
(f) 1 real buys $0.565131 or, equivalently, 1 real = $0.565131.
   1 real buys $0.570776 or, equivalently, 1 real = $0.570776.
(h) During the period between the early and late points in time, the krone has appreciated by 0.39% against the dollar.

4. (a) 7 times 2 is first computed and the result is added to 5 to give $14 + 5 = 19$.
   (b) 2 times 3 is first computed and the result is subtracted from 8 to give $8 - 6 = 2$.

5. (a) First 15 is divided by 5; the result is multiplied by 2 to get 6.
   (b) First 10 is multiplied by 3; the result is divided by 2 to get 15.

6. (a) First 2^3 is computed. The result, 8, is multiplied by 7 to get 56, which is added to 5 to get 61.

7. The multiplier, -1, is used to ensure that the negative sign is not squared. The expression, $-x^2$, means that the x-value is squared and then preceded by a minus sign.

10. Excel raised 49 to the first power and then divided the result by 2.

11. (a) Let L = late value, E = early value; then
$$\frac{L-E}{E} = \frac{L}{E} - \frac{E}{E} = \frac{L}{E} - 1.$$

(b) $\dfrac{30-24}{24} = \dfrac{6}{24} = \dfrac{1}{4} = 0.25 = 25\%$; $\dfrac{30}{24} - 1 = \dfrac{5}{4} - 1 = 1.25 - 1 = 0.25 = 25\%$.

(c) $\dfrac{20-50}{50} = \dfrac{-30}{50} = -0.60 = -60\%$; $\dfrac{20}{50} - 1 = 0.40 - 1 = -0.60 = -60\%$.

CHAPTER 1

Section 1-1

1. (a) and (b) The table and graph appear in the following spreadsheet.

x	y = 4x + 7
0	7
1	11
2	15
3	19
4	23
5	27

(c) The y-intercept is 7.
(d) For every unit increase in x, the y-value increases by 4.
(e) For every unit decrease in x, the y-value decreases by 4.

2. (a) and (b) The table and graph are given in the following spreadsheet.
(c) The y-intercept is 3.
(d) For every unit increase in x, the y-value increases by 4.
(e) For every unit decrease in x, the y-value decreases by 4.

X	y = -2x+3						
0	3						
1	1						
2	-1						
3	-3						
4	-5						
5	-7						

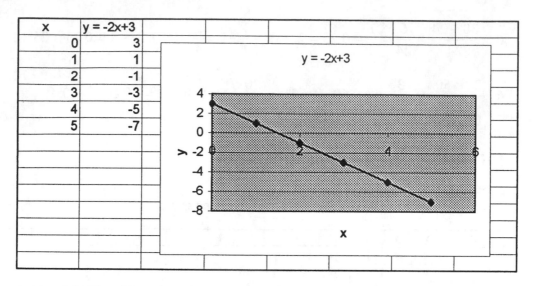

4. (a) and (b) The table and graph are given in the following spreadsheet.

X	y=4x	y=4x+3	y=4x-3			
0	0	3	-3			
1	4	7	1			
2	8	11	5			
3	12	15	9			
4	16	19	13			
5	20	23	17			

4. (c) The y-intercept for $y = 4x + 3$ is 3; for $y = 4x - 3$, it is –3; for $y = 4x$, it is 0.
   (d) $y = 4x$ passes through the origin.

6.

x	y = 2x	y = 5x
0	0	0
1	2	5
2	4	10
3	6	15
4	8	20
5	10	25

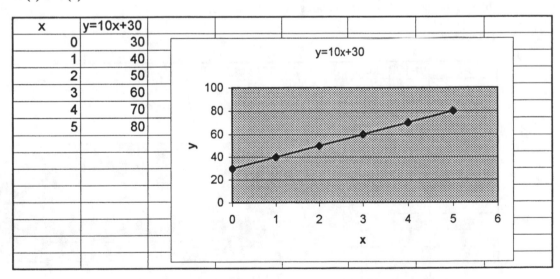

The steeper line has the greater slope. In this case, $y = 5x$ is the steeper line.

8. (a) and (b)

x	y=10x+30
0	30
1	40
2	50
3	60
4	70
5	80

(c) The y-intercept is 30.   (d) Fixed cost   (e) The fixed cost is the cost of producing 0 units. (f) For each unit increase in x, y increases by 10. (g) Each additional unit costs \$10 to produce. This is also called the variable cost per unit or, equivalently, the unit variable cost.

10. (a) (0.10)(8000) = $800 (b) $800 (c) $y = 8000 + 800x$
    (d) and (e)

x	y=8000+800x
0	8000
1	8800
2	9600
3	10400
4	11200
5	12000

(f) The y-intercept is 8000; it is called the initial investment or the initial value of the investment.
(g) For each unit increase in x, y increases by 800.
(h) The investment's value increases by $800 each year.

11. (a) and (b)

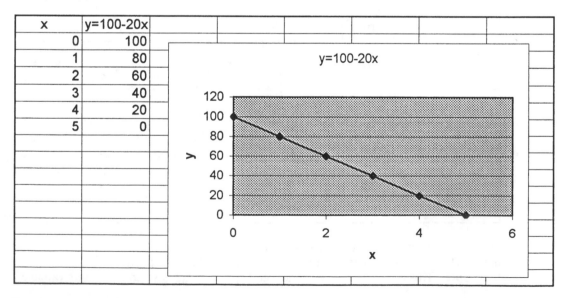

x	y=100-20x
0	100
1	80
2	60
3	40
4	20
5	0

(c) The y-intercept is 100. (d) For each unit increase in x, y decreases by 20.

(e) Inventory level decreases by 20 units per day.

Section 1-2

1. (a) $P(x) = R(x) - C(x)$

$$= 9x - (5x + 28)$$

$$= 4x - 28$$

(b) and (d)

x	R = 9x	C = 5x+28	P = R - C				
0	0	28	-28				
1	9	33	-24				
2	18	38	-20				
3	27	43	-16				
4	36	48	-12				
5	45	53	-8				
6	54	58	-4				
7	63	63	0				
8	72	68	4				
9	81	73	8				
10	90	78	12				
11	99	83	16				
12	108	88	20				

R=9x C=5x+28 P=R-C

(c) Because $P = R - C$, which, as we can see from part (a), results in a formula that gives profit in terms of x. (f) Observing the result of part (a), note that the negative fixed cost becomes the y-intercept of the profit function.

(g) Again, as we can see from part (a), the slope of the profit function equals the slope of the revenue function minus the slope of the cost function.

(h) The break-even point is where $R = C$. Hence,

$$9x = 5x + 28$$
$$4x = 28$$
$$x = 7$$

(i) Because the break-even point is where sales revenue equals cost or, equivalently, where profit equals 0. Profit is 0 at the x-intercept of the profit function.

3. (b) Given the profit functions $P(x) = 6x - 300$ and $P(x) = 6x - 299$, note that decreasing the fixed cost by \$1 means that the number subtracted from $6x$ is one less than the previous value so that the resulting profit is \$1 more.

5. (e)

$$P(x) = 6x - 360 \qquad\qquad\qquad P(x) = 6x - 240$$
$$0 = 6x - 360 \qquad\qquad\qquad 0 = 6x - 240$$
$$-6x = -360 \qquad\qquad\qquad -6x = -240$$
$$x = 60 \, BreakEvenPoint \qquad x = 40 \, BreakEvenPoint$$

(f) Yes

Section 1-3
1. (a) $x = 2$ (b) $y = 5$ 3. (a) $x = 7.2$ (b) $y = 1$
5. (a) $p = 21$ (b) $q = 17.33$ 9. (a) $p = 4.5$ (b) $q = 12$

Section 1-4
1. (a) Trendline: $y = 2.8333x + 3$ (b) $y = 2.8333(3.5) + 3$
$$= 12.91655$$

4. (a) Trendline: $y = 4x + 3$ (b) Each additional \$thousand spent on advertising increases monthly sales by \$4 thousand. (c) Yes, because the slope is nonzero. In other words, advertising expenditures do appear to have an effect on sales.

5. (a) Trendline is the horizontal line $y = 4$. (b) Slope is 0 which means that each additional \$thousand spent on advertising results in no change in monthly sales.
(c) No. The zero slope means that advertising expenditures do not appear to have an effect on sales.

6. (a) Trendline is the horizontal line $y = 3$. (b) Slope is 0. (c) No. The zero slope means that there is no relationship between rod length and time. In other words, the passage of time does not appear to have an effect on rod length.

CHAPTER 2

Section 2-1

1. (c) Yes.   2. (c) Yes.   3. Graph should resemble that of Spreadsheet 2-1.   4. (c) Yes.
5. Graph should resemble that of Spreadsheet 2-2.

6. (d) The thinner graph is that of $y = 4x^2$. Multiplying by 4 bends the graph thinner.

7. (d) The thinner graph is that of $y = x^2$. Multiplying x^2 by 1/2 bends the graph wider.

12.

x	y=x^2	y=x^3					
0	0	0					
0.1	0.01	0.001					
0.2	0.04	0.008					
0.3	0.09	0.027					
0.4	0.16	0.064					
0.5	0.25	0.125					
0.6	0.36	0.216					
0.7	0.49	0.343					
0.8	0.64	0.512					
0.9	0.81	0.729					
1	1	1					

$y = x^2$ vs $y = x^3$

(c) The lower graph is that of $y = x^3$.

Section 2-2

1. (d) The higher graph is that of $y = x^2 + 7$. (e) The y-coordinates of $y = x^2 + 7$ are 7 greater than those of $y = x^2$.

6. The graph of $y = (x + 3)^2$ is that of $y = x^2$ shifted horizontally 3 units to the left.

Section 2-3

1.

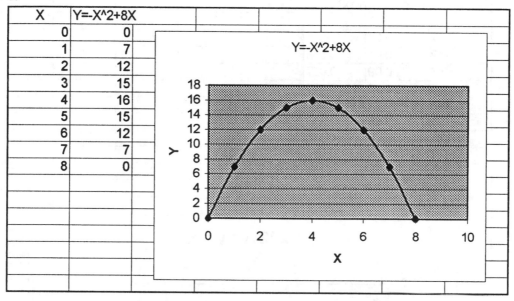

X	Y=-X^2+8X
0	0
1	7
2	12
3	15
4	16
5	15
6	12
7	7
8	0

(d) $0 = x(-x + 8)$ Setting each factor equal to 0 gives $x = 0$ or $x = 8$.

(g) The y-intercept is 0.

(h) Substituting 0 for x into $y = x(-x + 8)$ gives $y = 0(-0 + 8) = 0$.

4. (a) For the form $y = ax^2 + bx$, substituting $x = 0$ gives $y = a(0)^2 + b(0) = 0$, which means that the origin (0, 0) is a point on the graph.

(b) Factoring out x, the form $y = ax^2 + bx$ is written as $y = x(ax + b)$. Setting $y = 0$ and solving for x gives the x-intercepts $x = 0$ and $x = -b/a$ as illustrated below.

$$x = 0 \qquad ax + b = 0$$
$$ax = -b$$
$$x = \frac{-b}{a}$$

(c) The x-coordinate of the vertex is midway between the x-intercepts, 0 and $-b/a$, and is therefore one-half the nonzero x-intercept, or $\dfrac{1}{2} \cdot \dfrac{-b}{a} = \dfrac{-b}{2a}$.

(d) Substituting $x = 0$ into the equation $y = ax^2 + bx$ gives $y = 0$.

9.

X	y=-x^2+4x	y=-x^2+4x+12				
-2	-12	0				
-1	-5	7				
0	0	12				
1	3	15				
2	4	16				
3	3	15				
4	0	12				
5	-5	7				

(b) Because, due to a vertical shift, one graph is 12 units higher than the other so that they both have the *same axis of symmetry* and therefore the x-coordinate of the vertex is the same for both graphs.   (c) 12.

Section 2-4

1. (c) $R(x) = px$

$$= (40 - 2x)x$$

$$= 40x - 2x^2$$

Another formula is =40*A2-2*A2^2 or =-2*A2^2+40*A2.

CHAPTER 3

Section 3-1

1. (b) Graph should resemble that of $y = \dfrac{1}{x}$. (c) The y-axis, whose equation is x = 0, is the vertical asymptote; the x-axis, whose equation is y = 0, is the horizontal asymptote. (d) Yes.

2. (b) Graph should resemble that of $y = \dfrac{1}{x^2}$. (c) The y-axis, whose equation is $x = 0$, is the vertical asymptote; the x-axis, whose equation is $y = 0$, is the horizontal asymptote. (d) Yes.

3. (c) $y = \dfrac{5}{x}$ 5. (c) Negative sign turns the original graph upside down.

8. (a) Another formula is =5*A2+2/A2.

11. (d) The y-values of $y = 5 + 7x + \dfrac{3}{x}$ are 5 greater than those of $y = 7x + \dfrac{3}{x}$.

Section 3-2

1. (b) $A(x) = \dfrac{5x+3}{x}$; $A(x) = 5 + \dfrac{3}{x}$. The equation $A(x) = 5 + \dfrac{3}{x}$ reveals the horizontal asymptote, whose equation is y = 5. (d) 5; Yes; $A(x) = 5 + \dfrac{3}{x}$

3. (a) $A(x) = \dfrac{x^2 - 60x + 1600}{x}$; $A(x) = x - 60 + \dfrac{1600}{x}$
(d) Oblique asymptote is y = x.

(e) Minimum average cost per unit is $20, and 40 units must be produced in order to achieve this minimum.

5. (c) Oblique asymptote is y = 8x. (d) Minimum inventory cost is $16,000; it occurs at an order size of 1000 units.

CHAPTER 4

Section 4-1

2. (a) $M + W$ (b) $W + F$ (c) $M + W + F$ (d) $M + 2W$ (e) $2M - F$ (f) $M + W - F$

4. The answer is given by matrix XC in the following spreadsheet.

		Matrix X					Matrix C
	Product A	Product B	Product C				UnitVarCo
April	200	500	300		Product A		30
May	400	200	600		Product B		20
June	600	800	900		Product C		50
			Matrix XC				
		April	31000				
		May	46000				
		June	79000				

5. *TV*   6. AN   7. AN

Section 4-2

1. (b) The solution is given as matrix A^-1B in the following spreadsheet.

MATRIX A			MATRIX B	
1	2		40	
5	4		140	
MATRIX A^-1			SOLUTIO	A^-1B
-0.66667	0.333333			20
0.833333	-0.16667			10

(c) The solution is given as matrix A^-1B in the following spreadsheet.

MATRIX A			MATRIX B	
1	2		31	
5	4		107	
MATRIX A^-1			SOLUTIO	A^-1B
-0.66667	0.333333			15
0.833333	-0.16667			8

(d) The answer is given as matrix AX in the following spreadsheet.

MATRIX A			MATRIX X		MATR AX
1	2		30		60
5	4		15		210

2. (b) The solution is given as matrix A^-1B in the following spreadsheet.

	MATRIX A			MATRIX B	
1	3	5		31000	
5	6	8		69000	
1	1	1		11000	
	MATRIX A^-1			MATRIX	A^-1B
-1	1	-3			5000
1.5	-2	8.5			2000
-0.5	1	-4.5			4000
	MATRIX	X		MATRIX	AX
		6000			50000
		3000			104000
		7000			16000

(c) The results are derived from matrix AX given in the above spreadsheet. Specifically, the revised total profit is $50,000; the revised total cost is 20,000 + 104,000 = $124,000; the revised total number of units produced is 16,000.

5. (c) Week 1: A = 20, B = 30, C = 40, D = 10. (d) Week 2: A = 40, B = 20, C = 10, D = 20.

Section 4-3

3. (a) The SOLUTION MATRIX is given in the following spreadsheet.

	MATRIX A			IDENTITY MATRIX I	
0.1	0.2	0.2		1	0	0
0.2	0.1	0.2		0	1	0
0.2	0.2	0.1		0	0	1
MATRIX I - A				MATRIX (I - A)^-1	
0.9	-0.2	-0.2		1.272727	0.363636	0.363636
-0.2	0.9	-0.2		0.363636	1.272727	0.363636
-0.2	-0.2	0.9		0.363636	0.363636	1.272727
	MATRIX D			SOLUTION MATRIX X=(I-A)^-1	
	100			265.4545	
	200			356.3636	
	180			338.1818	
	INTERNAL CONSUMPTION AX				
		165.4545			
		156.3636			
		158.1818			

(b) The solution is given in the above spreadsheet as INTERNAL CONSUMPTION AX.

CHAPTER 5

1. $x_1 = 0$, $x_2 = 0$, $x_3 = 750$. Maximum profit = $45,000. Slack in first constraint is 0; slack in second constraint is 450.

2. $x_1 = 1000$, $x_2 = 0$, $x_3 = 0$. Maximum profit = $600,000. Slack in first constraint is 3000; slack in second constraint is 0.

4. $x_1 = 4$, $x_2 = 0$, $x_3 = 2.5$. Minimum calories = 4700. Surplus in first constraint is 0; surplus in second constraint is 0; surplus in third constraint is 5.5.

6. $x_1 = 244{,}444.47$, $x_2 = 100{,}000$, $x_3 = 205{,}555.55$, $x_4 50{,}000$. Maximum annual return = $58,583.34.

CHAPTER 6

Section 6-1

1. (b) At simple interest: $2500; at compound interest: $2552.56. (c) At simple interest: $3000; at compound interest: $3257.79. (d) Linear graph represents simple interest, while the curved graph represents compound interest.

2. (a) A table and graph are given in the spreadsheet on the next page.
(b) $y = 5(1.60)^x$; yes.

11. 39.42% compounded quarterly.　　　　12. 35.12% compounded annually.

13. 22.51 half-year time periods.

15. (a) 19.9 years.　(b) 10.4 years.　(c) 8.5 years.　(d) decrease.

16. $31,511.50　　　19. (a) $24,718.47　　　20. (a) $13,136.71 (b) $18,455.67 (c) 40.49%

Section 6-2

1. $111,109.57　　　3. $103,699.03　　4. 11.57%　　5. 5.17%

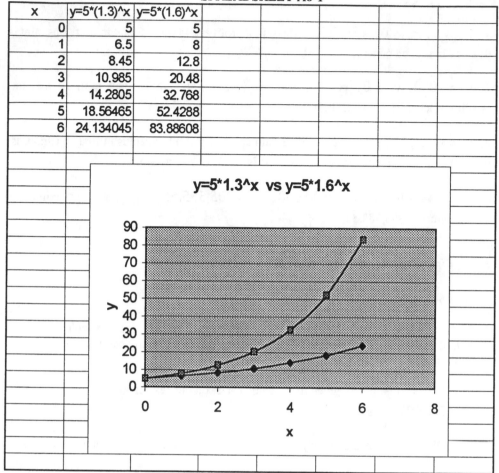

x	y=5*(1.3)^x	y=5*(1.6)^x				
0	5	5				
1	6.5	8				
2	8.45	12.8				
3	10.985	20.48				
4	14.2805	32.768				
5	18.56465	52.4288				
6	24.134045	83.88608				

Section 6-3
1. (a) $18,253.12 2. (a) $18,618.18 5. (b) 26.84% compounded monthly

8. 22.5 half-year time periods. 12. (a) $2159.20 (b) $2519.07

Section 6-4
1. (a) $11,348.36 (c) $7176.68   2. (a) $11,575.32   (c) $7356.10

3.   Note that for each exercise,

$$\frac{Pr esentValueAnnuityDue}{Pr esentValueOrdinaryAnnuity} = 1+i.$$

Solving this equation for *PresentValueAnnuityDue*, we multiply both sides by the denominator to get

$$Pr\,esentValueAnnuityDue = (1+i)\,Pr\,esentValueOrdinaryAnnuity\,.$$

10. (d) Decrease

Section 6-5
1. (b) Decreases

3. Monthly payment is $665.30. (a) Mortgage balance after 10 years is $85,812.82; mortgage balance after 15 years is $74,019.66; mortgage balance after 20 years is $57,301.38.
(b) $Interest = (360)(665.30) - 100,000 = \$139,508$.

5. Monthly payment is $898.83.
(b) $Interest = (180)(898.83) - 100,000 = \$61,789.40$.
Comparing this against the $139,508 interest paid with the 30-year mortgage results in a saving of $\$139,508 - 61,789.40 = \$77,718.60$.

7. (c) We will compare the total amounts paid with and without the additional $1000 payment.

Without additional $1000 payment:

Sum of payments made = (360)(665.30)= $239,508.

With additional $1000 payment:

Sum of payments made:

349 payments @ $665.30 = (349)(665.30) = $232,189.70

plus additional $1000 payment   +1000.00

minus $28.83 overpayment with last payment   -28.83

                                            $233,160.87
Saving = $239,508 - 233,160.87 = $6347.13. Thus, the additional $1000 payment results in a saving of $6347.13 over the life of the mortgage.

Section 6-6
1. (a) The monthly lease payment is $567.06.

2. (b) As the capitalized cost increases, the monthly lease payment increases.

4. (b) As the residual value increases, the monthly lease payment decreases.

6. (a) Because the quantity $C - V(1+i)^{-n}$ becomes the present value of an annuity whose payment is to be determined. Another way of looking at this is to note that the quantity $C - V(1+i)^{-n}$ becomes the amount of a loan whose monthly payment is to be determined. Thus, the smaller the loan amount, the smaller the payment.

8. (a) A decrease in the MSRP decreases the capitalized cost.
(b) An increase in the trade-in allowance decreases the capitalized cost.
(c) An increase in the deposit decreases the capitalized cost.

CHAPTER 7

Section 7-1
2. (a) Expected value for Gambling Machine 1 is 2.02. Expected value for Gambling Machine 2 is 3.98.
(b) For Gambling Machine 1, the larger probabilities occur at smaller x-values, whereas for Gambling Machine 2, the larger probabilities occur at larger x-values. Therefore, the expected value for Gambling Machine 2 is greater than for Gambling Machine 1.
(c) Casino should charge $3.02 per play.
(d) Expected profit per play is $2.02.

4. (b) Expected value (Gain) is $480. (c) No. Expected value is realized in the long run.

Section 7-2
1. (a) The second distribution appears to have greater variability because the larger probabilities are more spread out and so not as concentrated as those of the first distribution.
(b) Variance for the first distribution is 0.7; variance for the second distribution is 1.6.
(c) The outer x-values, $x = 1$ and $x = 5$, provide greater contributions to the variance.

4. (a) Whoever is inspecting the product is passing the metal rods falling just below the lower specification limit. In other words, the rods falling just below the lower specification limit, although defective, are being classified as acceptable because the inspector is (most likely) protecting fellow workers.

Section 7-3
1. (d) Larger probabilities occur at smaller x-values because the probability, p, of getting a success on an individual trial is small so there are likely to be fewer successes.

5. (a) The binomial distribution is given in the following spreadsheet.

Binomial Distribution n = 10, p = .6			
	X	P(X)	
	0	0.000105	
	1	0.001573	
	2	0.010617	
	3	0.042467	
	4	0.111477	
	5	0.200658	
	6	0.250823	
	7	0.214991	
	8	0.120932	
	9	0.040311	
	10	0.006047	
		1	

(c) Study the above distribution and note that the probability of getting 2 or fewer viewers buying the product is very small (i.e., $0.000105 + 0.001573 + 0.010617 = 0.012295$). This implies that if $p = 0.6$, this event (only 2 viewers buying the product) is not likely to happen. The fact that it did happen provides contradictory evidence that allows us to question whether $p = 0.6$.

6. (a) See the following spreadsheet. (b) $p = 0.5$ (c) $p = 0.1$ and $p = 0.9$.

n	p	Variance
10	0.1	0.9
10	0.2	1.6
10	0.3	2.1
10	0.4	2.4
10	0.5	2.5
10	0.6	2.4
10	0.7	2.1
10	0.8	1.6
10	0.9	0.9

Section 7-4
1. (a) 0.00135   (b) 0.00621   (e) 0.5 (g) 0.81594   3. (a) 0.191462 4. (a) 0.49702

5. (a) 0.009245   6. (a) 332.6703   7. (c) 0.037538   (d) 0.5   8. (a) –0.5244

10. (a) $x \le 25$ (b) $x < 30$ (c) $x > 30$ (d) 0.5 (e) 0.841345

(f) 0.158655   (g) 0.0227

CHAPTER 8

Section 8-1

1. Limit equals 12.

2. The function values (in the spreadsheet below) are approaching 0.5. This indicates that the limit is 0.5.

h	$((1+h)^{.5}-1)/h$			h	$((1+h)^{.5}-1)/h$
-0.5	0.585786438			0.5	0.449489743
-0.4	0.563508327			0.4	0.458039892
-0.3	0.544466578			0.3	0.467251417
-0.2	0.527864045			0.2	0.477225575
-0.1	0.513167019			0.1	0.488088482
-0.01	0.501256289			0.01	0.498756211
-0.001	0.500125063			0.001	0.499875062
-0.0001	0.500012501			0.0001	0.499987501
-0.00001	0.50000125			0.00001	0.49999875
-0.000001	0.500000125			0.000001	0.499999875

3. As x approaches 0 from the *left* (negative x-values in the spreadsheet), the y-values are -1; as x approaches 0 from the *right* (positive x-values in the spreadsheet), the y-values are 1. The fact that the y-values *do not approach the same number from both directions* means that the *limit does not exist* at $x = 0$.

4. For the spreadsheet below, as we move down the $1/x$ column corresponding to the negative x-values, the y-values differ from each other by larger and larger amounts as the x-values approach 0. The same occurs for the $1/x$ values corresponding to the positive x-values. In other words, the y-values increase without bound as x approaches 0. This means that this function has <u>no limit</u> as x approaches 0.

x	1/x			x	1/x
-0.1	-10			0.1	10
-0.01	-100			0.01	100
-0.001	-1000			0.001	1000
-0.0001	-10000			0.0001	10000
-0.00001	-100000			0.00001	100000
-0.000001	-1000000			0.000001	1000000

Section 8-2

1. The **[f(x + h)-f(x)]/h** columns in Spreadsheets A8-3 and A8-4 indicate the instantaneous rate of change to be 80.

242

SPREADSHEET A8-3

x	h	x+h	f(x+h)	f(x)	[f(x+h)-f(x)]/h
4	0.1	4.1	168.1	160	81
4	0.01	4.01	160.801	160	80.1
4	0.001	4.001	160.08	160	80.01
4	0.0001	4.0001	160.008	160	80.001
4	0.00001	4.00001	160.0008	160	80.0001

SPREADSHEET A8-4

x	h	x+h	f(x+h)	f(x)	[f(x+h)-f(x)]/h
4	-0.1	3.9	152.1	160	79
4	-0.01	3.99	159.201	160	79.9
4	-0.001	3.999	159.92	160	79.99
4	-0.0001	3.9999	159.992	160	79.999
4	-0.00001	3.99999	159.9992	160	79.9999

5. Instantaneous rate of change is –0.25.   7. Instantaneous rate of change is 8.

Section 8-3

4. Observing the difference quotients, $[f(x + h) - f(x)]/h$, as we move down the column (either in Spreadsheet A8-5 or Spreadsheet A8-6), note that the successive difference quotient values differ from each other by larger and larger amounts so that they appear to be increasing without bound. This means that the function is <u>not differentiable</u> at the indicated x-value—in this case, $x = 0$.

SPREADSHEET A8-5

x	h	x+h	f(x+h)	f(x)	[f(x+h)-f(x)]/h
0	0.1	0.1	0.464159	0	4.641588834
0	0.01	0.01	0.215443	0	21.5443469
0	0.001	0.001	0.1	0	100
0	0.0001	0.0001	0.046416	0	464.1588834
0	0.00001	0.00001	0.021544	0	2154.43469

SPREADSHEET A8-6

x	h	x+h	f(x+h)	f(x)	[f(x+h)-f(x)]/h
0	-0.1	-0.1	-0.46416	0	4.641588834
0	-0.01	-0.01	-0.21544	0	21.5443469
0	-0.001	-0.001	-0.1	0	100
0	-0.0001	-0.0001	-0.04642	0	464.1588834
0	-0.00001	-0.00001	-0.02154	0	2154.43469

7. The difference quotient columns in Spreadsheets A8-7 and A8-8 both indicate that the function is differentiable with the value of the derivative equal to 1.

SPREADSHEET A8-7

x	h	x+h	f(x+h)	f(x)	[f(x+h)-f(x)]/h
2	0.1	2.1	2.1	2	1
2	0.01	2.01	2.01	2	1
2	0.001	2.001	2.001	2	1
2	0.0001	2.0001	2.0001	2	1
2	0.00001	2.00001	2.00001	2	1

SPREADSHEET A8-8

x	h	x+h	f(x+h)	f(x)	[f(x+h)-f(x)]/h
2	-0.1	1.9	1.9	2	1
2	-0.01	1.99	1.99	2	1
2	-0.001	1.999	1.999	2	1
2	-0.0001	1.9999	1.9999	2	1
2	-0.00001	1.99999	1.99999	2	1

8. In contrast to Spreadsheets A8-7 and A8-8 of Exercise 7, the fact that the difference quotient columns in Spreadsheets A8-9 and A8-10 *do not both approach the same number* indicates that the function is <u>not differentiable</u> at the indicated x-value—in this case, $x = 0$. In other words, the difference quotients both to the *left* and to the *right* of $x = 0$ should be approaching the same number.   The fact that they are <u>not</u> indicates that the derivative does <u>not</u> exist.

SPREADSHEET A8-9

x	h	x+h	f(x+h)	f(x)	[f(x+h)-f(x)]/h
0	0.1	0.1	0.1	0	1
0	0.01	0.01	0.01	0	1
0	0.001	0.001	0.001	0	1
0	0.0001	0.0001	0.0001	0	1
0	0.00001	0.00001	0.00001	0	1

SPREADSHEET A8-10

x	h	x+h	f(x+h)	f(x)	[f(x+h)-f(x)]/h
0	-0.1	-0.1	0.1	0	-1
0	-0.01	-0.01	0.01	0	-1
0	-0.001	-0.001	0.001	0	-1
0	-0.0001	-0.0001	0.0001	0	-1
0	-0.00001	-0.00001	0.00001	0	-1

CHAPTER 9

Section 9-1

2. See the following spreadsheet.

x	y=8*1.7^-x
0	8
1	4.705882
2	2.768166
3	1.628333
4	0.957843
5	0.563437
6	0.331434

6. (a) and (b) See Spreadsheet A9-1.

SPREADSHEET A9-1

x	y=e^x	y=e^1.5x
-2	0.135335	0.049787
-1	0.367879	0.22313
0	1	1
1	2.718282	4.481689
2	7.389056	20.08554
3	20.08554	90.01713

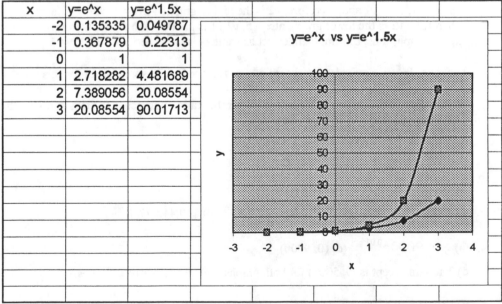

6. (c) Observing the table of x- and y-values of Spreadsheet A9-1, note that the y-intercept for both graphs is (0, 1).
(d) Observing the table of x- and y-values of Spreadsheet A9-1, note that for $x > 0$, the smaller y-values belong to $y = e^x$. Therefore, it is the lower graph on the interval $x > 0$. For $x < 0$, the smaller y-values belong to $y = e^{1.5x}$.
(e) 0. Accordingly, the x-axis ($y = 0$) is a horizontal asymptote for both graphs.   (f) No.

7. (b) 128; $1.28   (c) 16,384; $163.84   (d) 1,073,741,824; $10,737,418.24
(e) $y = 5 \cdot 2^x$ (g) 5,368,709,120; $53,687,091.20

9. (a) $y = 7 \cdot 3^x$ 13. (a) $y = 8000 \cdot 3^{-x}$ 16. $y = 30 \cdot 1.12^x$

17. (c) The y-intercept is (0, 5) for both graphs.   (d) $y = 5 \cdot 1.6^x$ (e) Yes.

18. (b) Data set 1 exhibits exponential *growth* because the *ratios are greater than 1*. Data Set 2 exhibits exponential *decay* because the *ratios are less than 1*. Data set 3 exhibits exponential *growth* because the *ratios are greater than 1*.

20. The limit is 1.

22. (b) At $x = 1$, $y1$ is the higher graph because its y-coordinate is greater than that of $y2$. At $x = 5$, $y1$ is the lower graph because its y-coordinate is smaller than that of $y2$. Therefore, the graphs intersect at some point between $x = 1$ and $x = 5$.

(c) When $y2/y1$ is less than 1, $y2 < y1$ and, therefore, $y2$ is the lower graph; when $y2/y1$ is greater than 1, $y2 > y1$ and, therefore, $y2$ is the higher graph. At $x = 1$, $y2/y1$ is less than 1 and, therefore, $y2$ is the lower graph; at $x = 5$, $y2/y1$ is greater than 1 and, therefore, $y2$ is the higher graph; thus, the graphs intersect as some point between $x = 1$ and $x = 5$.

(f) The ratio $\dfrac{y2}{y1}$ is equivalent to the ratio $\dfrac{x^p}{e^x}$ for $p = 6$. The spreadsheet for this problem indicates that as x gets larger and larger, the ratio approaches 0. This supports the summary given at the end of Exercise 21.

Section 9-2

2. (a) $5809.17   (b) $12,928.60   (c) $6087.85   (d) $10,442.77

4. (a) $y = 9000e^{0.07x}$ (d) (0, 9000)

5. (d) The y-intercept is (0, 5000) for both graphs.   (e) $y = 5000e^{0.09x}$; yes.

8. (b) $19,711.40   9. $r = 0.211824 = 21.18\%$ 10. $r = 0.150408 = 15.04\%$
12. $t = 27.72589$ years   15. $t = 11.54595$ years   19. $t = 10.98687$ years

1. $x = 1.60937$ 5. $x = 4.49981$ 6. $x = \ln 5$ 10. $x = \ln 90$ 11. $x = \ln y$ 12. $x = e^y$
14. (d) $5.010635 - 4.60517 = 0.405465$ (e) $5.298317 - 5.010635 = 0.287682$

15. (a) and (b) See the following spreadsheet.

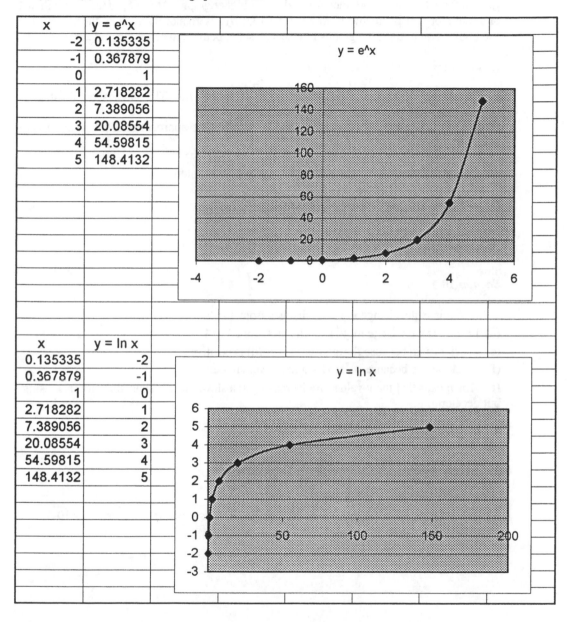

x	y = e^x
-2	0.135335
-1	0.367879
0	1
1	2.718282
2	7.389056
3	20.08554
4	54.59815
5	148.4132

x	y = ln x
0.135335	-2
0.367879	-1
1	0
2.718282	1
7.389056	2
20.08554	3
54.59815	4
148.4132	5

15. (c) Because $y = \ln x$ means the same as $x = e^y$. (d) Interchanging the x- and y-
coordinates of $y = e^x$ will give the graph of $y = \ln x$, and vice versa.

17. (c) At $x = 5$, $y1$ is the lower graph because its y-coordinate is smaller than that of $y2$. At $x = 10$, $y1$ is the higher graph because its y-coordinate is larger than that of $y2$. Therefore, the graphs intersect at some point between $x = 5$ and $x = 10$.

(d) When $y1/y2$ is less than 1, $y1 < y2$ and, therefore, $y1$ is the lower graph; when $y1/y2$ is greater than 1, $y1 > y2$ and, therefore, $y1$ is the higher graph. At $x = 5$, $y1/y2$ is less than 1 and, therefore, $y1$ is the lower graph; at $x = 10$, $y1/y2$ is greater than 1 and, therefore, $y1$ is the higher graph; thus, the graphs intersect as some point between $x = 5$ and $x = 10$.

(f) $y1 = \ln x$

(g) The ratio $\dfrac{y1}{y2}$ is equivalent to the ratio $\dfrac{\ln x}{x^p}$ for $p = 1/3$. The spreadsheet for this problem indicates that as x gets larger and larger, the ratio approaches 0. This supports the summary given at the end of Exercise 16.

18. (a) $a = 5$, $b = 3$ (b) $\ln b = \ln 3 = 1.0986$ is the slope.

(c) $\ln a = \ln 5 = 1.6094$ is the y-intercept.

CHAPTER 10

Section 10-1

1. (b) 5; this is the change in y as x changes from 1 to 2.
(d) 13; this is the change in y' as x changes from 3 to 4.
(e) Accelerating; because the answer to part (d) is positive.
(f) Decelerating; because $y'' = -15$, a negative number.
(g) This means that the y-values are increasing at a slower rate. Thus, they are increasing but decelerating.

3. Increasing; decreasing.

4. Increasing; accelerating.   Decreasing; decelerating.

7. (a) and (c) are given in Spreadsheet A10-1.
(b) Decelerating; because the S''-values are negative as given in Spreadsheet A10-1.

SPREADSHEET A10-1

t	S= -16*t^2+192t	S'	S"				
0	0						
1	176	176					
2	320	144	-32				
3	432	112	-32				
4	512	80	-32				
5	560	48	-32				
6	576	16	-32				
7	560	-16	-32				
8	512	-48	-32				
9	432	-80	-32				
10	320	-112	-32				
11	176	-144	-32				
12	0	-176	-32				

S= -16*t^2+192t

Section 10-2

1. See Spreadsheet A10-2.

(a) f is increasing on the interval(s) where f' is positive. The table and graph of y' in Spreadsheet A10-2 show that these intervals are $x < -2$ and $x > 3$.

(b) f is decreasing on the interval(s) where f' is negative. The table and graph of y' in Spreadsheet A10-2 show that these intervals are $-2 < x < 3$.

(c) Local maximum at $x = -2$; local minimum at $x = 3$.

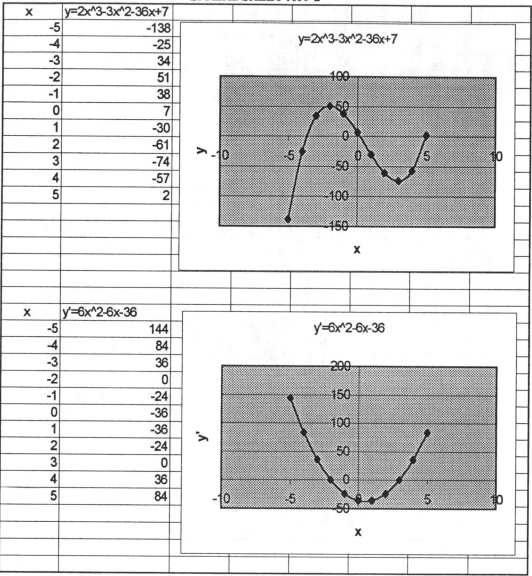

x	y=2x^3-3x^2-36x+7
-5	-138
-4	-25
-3	34
-2	51
-1	38
0	7
1	-30
2	-61
3	-74
4	-57
5	2

x	y'=6x^2-6x-36
-5	144
-4	84
-3	36
-2	0
-1	-24
0	-36
1	-36
2	-24
3	0
4	36
5	84

4. See Spreadsheet A10-3.

(a) $x < -3$ and $x > 3$ (b) $-3 < x < 3$

(c) Local maximum at $x = -3$ because the first derivative is positive to the left of and negative to the right of $x = -3$. This indicates that the function is increasing to the left and decreasing to the right of $x = -3$, which implies a local maximum at $x = -3$. Local minimum at $x = 3$ because the first derivative is negative to the left of and positive to the right of $x = 3$. This indicates that the function is decreasing to the left of and increasing to the right of $x = 3$, which implies a local minimum at $x = 3$.

SPREADSHEET A10-3

x	y' = x^2 -9				
-4	7				
-3	0				
-2	-5				
-1	-8				
0	-9				
1	-8				
2	-5				
3	0				
4	7				

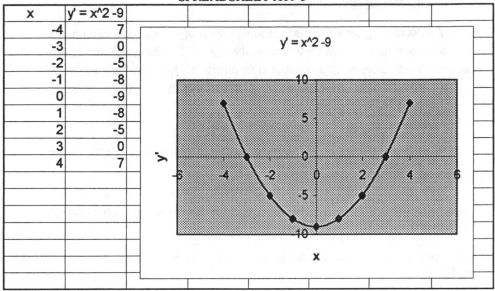

7. (a) $x < 0$ (b) $x > 0$ (c) At $x = 0$. This point is neither a local maximum nor a local minimum because the function is <u>not</u> defined here (at $x = 0$).

8. (a) $x > \frac{1}{2}$ (b) $x < \frac{1}{2}$ (c) Inflection point at $x = \frac{1}{2}$ because the second derivative changes sign here.

10. (a) $x < 0$ and $x > 4/3$ (b) $0 < x < 4/3$ (c) At $x = 0$ and $x = 4/3$; because the second derivative changes sign at these x-values.

11. (a) All values of x. (b) Nowhere. (c) Because the second derivative does not change sign at any x-value.

18. (a) f' is increasing at values of x where f'' is positive. Observing the graph of y'' at the bottom of Spreadsheet 6-6, note that the x-intercept of y'' is determined by setting $y'' = 0$ and solving for x to get $x = 54/12$, or, equivalently, 4.5. Thus, as we can see from the graph, y'' is positive on the interval $x < 4.5$, and therefore, f' is increasing on $x < 4.5$.
(b) f' is decreasing at values of x where f'' is negative. Observing the graph of y'' at the bottom of Spreadsheet 6-6, note that y'' is negative on the interval $x > 4.5$, and therefore, f' is decreasing on this interval.
(c) Just as a sign change in f' from positive to negative as we move in a left-to-right direction indicates a local maximum value of f, so does a sign change in f'' from positive to negative as we move in a left-to-right direction indicate a local maximum value of f'.

This occurs at $x = 4.5$. Equivalently, because f' is the rate of change of f, we know that f is increasing most rapidly at this point.

22. (a) f is decreasing for $x < 0$; f is increasing for $x > 0$; f has a local minimum at $x = 0$.
(b) f has a local minimum at $x = 0$ because the sign of f' changes from negative to positive at $x = 0$ as we move in a left-to-right direction. However, the graph of f comes to a sharp point (i.e., a cusp) at $x = 0$ because f' does not exist at $x = 0$.

Section 10-3

1. (a) $A(x) = x - 80 + \dfrac{3600}{x}$ (b) 60 units should be produced in order to minimize the average cost per unit; the minimum average cost is $40.

2. (a) 1000 units is the order size that minimizes the total annual inventory cost; the minimum inventory cost is $4000.

4. (a) 6 units should be sold in order to maximize profit. 5. A local minimum occurs at $x = -2$.

CHAPTER 11

Section 11-1

1. (a) 20 mph   (b) 50 mph   (c) 80 mph
(d) $20(2 - 1) + 50(4 - 2) = 20(1) + 50(2) = 120$ miles
(e) $20(2 - 1) + 50(6 - 2) + 80(7 - 6) = 20(1) + 50(4) + 80(1) = 300$ miles

2. (a) See the following spreadsheet.

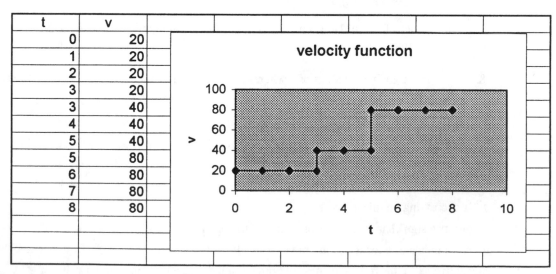

3. (a) $10 (b) $8 (c) $4 (d) $10(3 – 1) + 8(4 – 3) = 10(2) + 8(1) = 28
(e) $8(6 – 4) + 4(7 – 6) = 8(2) + 4(1) = 20 (f) $10(3 – 2) + 8(6 – 3) = 10(1) + 8(3) = 34

7. (a) 0.10 or 10% (b) 0.15 or 15% (c) 0.20 or 20% (d) $0.10(15) = 1.5 thousand = $1500
(f) $0.10(20) + 0.15(60 – 20) = 0.10(20) + 0.15(40) = 8 thousand = $8000
(g) $0.15(30 – 20) = 0.15(10) = 1.5 thousand = $1500
(i) $0.15(70 – 60) + 0.20(80 – 70) = 0.15(10) + 0.20(10) = 3.5 thousand = $3500

Section 11-2

1. (a) $x = 0.125, y = 0.015625$ (b) $x = 0.875, y = 0.765625$
(c) $x = 0, y = 0.015625$ (d) $x = 0.875, y = 1$
(e) Narrower interval results from thinner approximating rectangles.

3. (b) See the following spreadsheet.

LOWER APPROXIMATION				UPPER APPROXIMATION			
x	f(x)=x^3	dx	f(x)dx	x	f(x)=x^3	dx	f(x)dx
-1	-1	0.125		-1	-1	0.125	-0.125
-0.875	-0.66992	0.125	-0.08374	-0.875	-0.66992	0.125	-0.08374
-0.75	-0.42188	0.125	-0.05273	-0.75	-0.42188	0.125	-0.05273
-0.625	-0.24414	0.125	-0.03052	-0.625	-0.24414	0.125	-0.03052
-0.5	-0.125	0.125	-0.01563	-0.5	-0.125	0.125	-0.01563
-0.375	-0.05273	0.125	-0.00659	-0.375	-0.05273	0.125	-0.00659
-0.25	-0.01563	0.125	-0.00195	-0.25	-0.01563	0.125	-0.00195
-0.125	-0.00195	0.125	-0.00024	-0.125	-0.00195	0.125	-0.00024
0	0	0.125		0	0	0.125	
			-0.19141				-0.31641
Lower approximation			0.19141	Upper approximation			0.31641

(c) Observing this spreadsheet, note that the **f(x)dx** columns are negative because the *f(x)*-values are negative. Therefore, we take the absolute value of the sums of the **f(x)dx** columns to obtain the lower and upper approximations.

CHAPTER 12

Section 12-1

1. (a) $x = 4, y = 5$ 2. (a) $x = 10, y = 20$ 3. (a) $x = 5, y = 10$

6. (a) $x = 200, y = 500$ (b) Maximum profit is $1,660,000.

Section **12-2**

1. (a) $S = (10 - a - 2b)^2 + (15 - a - 5b)^2 + (25 - a - 8b)^2 + (30 - a - 9b)^2$
(b) $a = 3$, $b = 2.8333$ (c) $y = 3 + 2.8333x$

3. (a) $S = (8 - a - 2b)^2 + (10 - a - 3b)^2 + (7 - a - 2b)^2 + (15 - a - 5b)^2$
(b) $a = 2.5$, $b = 2.5$ (c) $y = 2.5 + 2.5x$ (d) $S = 0.5$

4. (a) $S = (7 - a - 3b)^2 + (5 - a - 2b)^2 + (10 - a - 4b)^2 + (6 - a - 3b)^2$
(b) $a = -0.5$, $b = 2.5$ (c) $y = -0.5 + 2.5x$ (d) $S = 1.5$

Section **12-3**

1. $x = 8$, $y = 10$; $f(8, 10) = 44$

2. $x = 5$, $y = 5$; $Z = f(5, 5) = 250$. Also, $\lambda = 50$, which means that a 1-unit increase in the right-hand-side constraint, 10, increases the optimal value of Z by 50 units. Accordingly, a 1-unit decrease in 10 decreases the optimal value of Z by 50 units.

3. $x = 10$, $y = -22$; $f(10, -22) = -252$.   7. $x = 20$, $y = 11.9$; $P = \$275.75$

INDEX